Electric Power Systems for Non-Electrical Engineers

This book explains the electrical power systems for non-electrical engineers and includes topics like electrical energy systems, electrical power systems structure, single-phase AC circuit fundamentals and three-phase systems, power system modeling, power system representation, power system operation, power flow analysis, economic operation of power systems, power system fault analysis, power system protection fundamentals, and so forth. Examples have been provided to clarify the description, and review questions are provided at the end of each chapter.

Features:

- Provides a simplified description of fundamentals of electrical energy systems and structure of electrical power systems for non-electrical engineers.
- Gives a detailed description of AC circuit fundamentals and three-phase systems.
- Describes power system modeling and power system representation.
- Covers power system operation, power flow analysis, and fundamentals of economic operation of power systems.
- Discusses power system fault analysis and fundamentals of power system protection with examples, and also includes renewable energy systems.

This book has been aimed at senior undergraduate and graduate students of non-electrical engineering background.

Electric Power Systems for Non-Electrical Engineers

Anup Tripathi

CRC Press is an imprint of the
Taylor & Francis Group, an **informa** business

Designed cover image: Dr.G.Schmitz, Coal-fired power station, Werdohl-Elverlingsen, Germany.

First edition published 2025
by CRC Press
2385 NW Executive Center Drive, Suite 320, Boca Raton FL 33431

and by CRC Press
4 Park Square, Milton Park, Abingdon, Oxon, OX14 4RN

CRC Press is an imprint of Taylor & Francis Group, LLC

ISBN: 9781032527789 (hbk)
ISBN: 9781032558103 (pbk)
ISBN: 9781003432340 (ebk)

DOI: 10.1201/9781003432340

Typeset in Times
by codeMantra

Dedicated to

My Parents

Contents

List of Figures

List of Tables

Preface

This book reference has been prepared after many years of work to help senior undergraduate and first-year graduate students of various engineering branches learn the principles of electrical power systems. Knowledge of only electrical engineering and engineering mathematics subjects taught in the first year of the undergraduate program in engineering is required to understand the text properly. The description in this book has been kept simple. Examples have been provided wherever required to clarify the description, and review questions are provided at the end of each chapter to help the student learn the chapter material more properly. This book reference will prepare the students of various engineering branches to understand more advanced texts on electric power systems.

This book has become possible due to the author's doctoral work in the area of mine electrical and power systems engineering at the University of Kentucky, Lexington, USA. It therefore mainly describes US electric power systems design and analysis. The description in the book reference is, however, applicable to electric power systems of any other country or region after few simple corrections in power system voltage levels and frequency.

The author wishes to thank his Electrical Engineering teachers at the University of Kentucky for their excellent teaching of electrical engineering subjects, numerous invaluable helps in research, and continued inspiration. The author further wishes to thank his parents for their affection and encouragement in writing the reference.

Author Biography

Dr. Anup Tripathi completed his Bachelor's degree in Mining Engineering with first division from the Indian Institute of Technology (ISM) Dhanbad, India, in 2000. He subsequently pursued his Master's degree in Mining Engineering from the Indian Institute of Technology, Kharagpur, India, from 2000 to 2002. He was the department topper in MTech program, and he was also among the best eight graduate students of IIT Kharagpur selected from all departments to complete his Master's degree project from Germany. He subsequently completed his PhD degree in Mining Engineering with specialization in the area of mine electrical and power systems engineering from the University of Kentucky, Lexington, USA, in 2007.

Dr. Anup Tripathi is presently working as a faculty of Mining Engineering at the National Institute of Technology Karnataka, Surathkal, India. His research interests include mine electrical and power systems engineering, renewable energy systems and environmental pollution control, mining systems optimization, and surface mining and reclamation. He has more than 50 national and international journal and conference publications.

List of Symbols

$\boldsymbol{a}$	Operator $1\angle 120°$
a	Transformer turns ratio
$\boldsymbol{A}, \boldsymbol{B}, \boldsymbol{C}, \boldsymbol{D}$	Transmission line constants
A	Area
C	Capacitance (F)
d	Diameter
$\boldsymbol{E}$	Generated E.M.F.
f	Frequency (Hz)
G	Conductance (Siemens)
H	Elevation (m)
i	Current (A)
$\boldsymbol{I}$	Current phasor (A)
$\boldsymbol{I}^*$	Conjugate of $\boldsymbol{I}$
l	Length
ln	Natural logarithm
N	Number of transformer winding turns
n	Speed of rotation (revolutions/min)
P	Power (W)
PF	Power factor
pu	Per-unit
Q	Reactive power (VAR), water flow rate (m^3/s)
R	Resistance (Ω)
$\boldsymbol{S}$	Complex power = $P + jQ$
S	Apparent power (VA)
s	Slip of induction motor
t	Time
v	Voltage (V)
$\boldsymbol{V}$	Voltage phasor (V)
X	Reactance (Ω)
$\boldsymbol{y}, \boldsymbol{Y}$	Admittance (Siemens)
$\boldsymbol{Z}$	Impedance (Ω)
η	Efficiency (%)
ρ	Resistivity (Ω.m), density (kg/m^3)
ε	Permittivity
μ	Permeability
θ	Impedance angle
$\boldsymbol{p}$	Propagation constant
γ	Admittance angle

δ	Bus voltage phase angle
ϕ	Magnetic flux
ϕ	Phase angle, phase quantity (as subscript)
ω	Angular frequency (rad/s)
Subscripts 0,1,2	Zero-, positive-, and negative-sequence symmetrical components

1 Fundamentals of Electrical Energy Systems

1.1 INTRODUCTION

Our ancestors lived without use of electricity in old times. In the modern world, however, it is nearly impossible to imagine survival without utilization of electricity. Humanity's most notable discovery probably has been the harnessing of electrical energy for performing different tasks. The high standards of living of the present times cannot be maintained without the availability of electrical energy systems.

The availability of reliable and efficient electrical energy systems has contributed to tremendous improvement in performing domestic activities and in industrial progress. Electricity is required for many domestic tasks such as cooking, cleaning, and washing. We are much dependent for proper living on devices such as television, computer, kitchen appliances, vacuum cleaner, and many others which require electricity to work. Electricity is needed for most of the technology in the commercial and industrial sectors. Education and transportation sectors also require electricity for carrying out various tasks.

Human societies require various forms of energy such as heat energy, light energy, and motive power to accomplish useful work. Technical advancement has made it possible to easily convert electrical energy to any other form of energy needed for human work. Electrical energy has therefore become extremely useful in the modern world. The availability of cheap and continuous supply of electrical energy is vital for industrial progress and for maintaining our social structures and high standards of living. The level of advancement of a nation is also determined from annual amount of electrical energy consumed per person.

This chapter describes the various phases of development of human civilization and the role of energy in bringing it to the present state. The usefulness of electrical energy and its many advantages over other forms of energy are described. A simplified arrangement for the generation of electrical energy from different energy sources and the efficiency of electrical power generation systems are also described. Finally, world electrical energy generation from different energy sources is described.

1.2 HUMAN CIVILIZATION AND ENERGY

Archaeological evidence indicate that humans have gradually learned to control and utilize different forms of energy available in nature for carrying out their tasks. They initially depended mainly on their muscular energy and solar energy, and later, with the passage of time, they learned to utilize animal power, wind energy, and water energy. After the Industrial Revolution started few hundred years ago, they built machines powered by coal, petroleum, and later electricity. Humans would never

DOI: 10.1201/9781003432340-1

have become civilized, and they would have forever remained uncivilized brutes if they had not learned how to control and utilize different forms of energy available in nature for performing their work [41].

Human civilizations mainly evolved in three stages. In the first stage, there were ancient hunters dependent on hunting wild animals and gathering other wild foods for their survival. In the second stage, humans started living as pastoral communities dependent on agriculture and farming for their sustenance. In the third stage, after the Industrial Revolution, humans started using machines powered by fossil fuels and later by electricity for their sustenance.

Ancient hunters lived in small, simple communities of few hundred individuals only. Activities such as obtaining food and shelter consumed most of their time, and there was very little time for other activities. After the development of agriculture, more reliable sources of food, other agricultural products, and energy became available. The societies which were semi-nomadic earlier then stated to have more permanent settlements.

The task of obtaining food for their sustenance still consumed most of the available time of early agricultural societies, and social interactions were very limited in general. Introduction of animal power for agricultural work reduced manpower requirement; people then had more time available for social interactions and activities, and social structures subsequently became more complex. Humans later learned to use wind power and water power for carrying out their tasks, and windmills and waterwheels were developed. These developments resulted in further availability of cheap and more abundant sources of energy in comparison with the past.

The use of water power and wind power resulted in the development of trade and transport between different communities. All this resulted in improved communication between various societies and more exchange of resources, ideas, and technical advancements. Development of sailing ships enhanced transport, trade, and communication between distant communities. Human activities then became more diverse, and specialized work disciplines such as farming, trading, shipping, and metalwork came into existence. Industry culture thus became much more commonplace.

The invention of steam engine led to the use of fossil fuels as an energy source. Later, the developments in electrical energy systems provided humans with tremendous power to control the surrounding environment and work toward improved social, political, and economic structure of the communities. All this resulted in more specialized work disciplines, and there was also greater social stability, improved living conditions, increased life expectancy, and improved supply and knowledge of healthy and balanced diet [41].

The high standards of living of the modern societies can be maintained and improved only with further industrial progress. It will therefore be necessary in future to develop new sources of energy which are abundant, economic, safe, reliable, and environment friendly.

1.3 ELECTRICAL ENERGY AND ITS USEFULNESS

A country's economic and technical advancement depends much on the availability of sufficient and large amount of energy. Without availability of energy, humans cannot perform many activities necessary for survival and daily living. Energy has played a central role in the advancement of human civilization and in bringing it to its present state. Abundance of energy supply has provided humanity with numerous benefits such as reduced manual work, increased agricultural and industrial output, availability of sufficient nourishing diet for all, and improved work environment and facilities. A nation's annual energy consumption per person is also directly related to the standard of living of its inhabitants.

Among the many different forms of energy available in nature, electric energy is the most useful. Due to its many useful qualities, it has found numerous applications in all sectors of the society, and its demand is also increasing steadily.

Electrical energy is generated mainly in power plants and instantaneously transmitted over long distances to the consumer using a network of power transmission and distribution lines. The transmission and distribution of electrical energy to supply it to the consumer is economic, efficient, and reliable. Electricity utilization at the consumer premises is also free from pollution. Presently available technology, however, does not allow the storage of large amounts of electricity, and the electricity that consumers consume was generated the moment before.

Electrical energy has become the most popular form of energy in comparison with all other forms of energy available in nature due to the following main reasons:

1. The generation of electrical energy in large quantities in power plants is cheap and economic in comparison with other forms of energy.
2. The generation of electrical energy is possible from renewable energy sources such as solar energy, wind energy, and hydro energy.
3. The generation of electrical energy without emission of greenhouse gases is possible, for example, in nuclear power plants and hydroelectric power plants.
4. Electrical energy can be transmitted over long distances by transmission and distribution lines to the place of utilization instantly, economically, efficiently, simply, and conveniently.
5. The availability of flexible metallic conductors allows electric energy to be delivered to any area of utilization.
6. The availability of suitable devices permits easy control, monitoring, and use of electrical energy.
7. At the place of utilization, electrical energy is pollution-free and free from smoke, fumes, and poisonous gases.
8. The availability of various devices makes it possible to easily convert electrical energy to other forms of energy such as heat energy, light energy, sound energy, and mechanical energy.
9. Noise levels produced by electrical equipment are generally much less.
10. Electrical energy can be stored in battery storage systems for later use.

1.4 SIMPLIFIED ARRANGEMENT FOR ELECTRICAL POWER GENERATION

There are different forms of energy available from various natural sources, including chemical energy from fossil fuels such as coal and natural gas, pressure head of water, and nuclear energy from radioactive substances such as uranium. Electrical energy generation is the process of converting energy present in different forms in nature into electrical energy.

Figure 1.1 shows a simplified arrangement used for conversion of different forms of energy into electrical energy. It mainly employs a prime mover such as a hydro-turbine driven by pressure of water or a steam turbine driven by steam produced at high pressure using heat energy obtained from burning of fossil fuel or heat produced during nuclear fission of radioactive substance. The mechanical energy of the prime mover is used to drive the three-phase synchronous electric generator for the generation of electrical power. The voltage level of the power generated is raised using grid transformers to supply power to the power transmission grid (network of very high voltage transmission lines) for transmission of electrical power. The electric power is transmitted close to the area of use and later distributed for utilization by loads at the consumer premises after proper reduction in voltage levels.

1.5 EFFICIENCY OF ELECTRIC POWER GENERATION SYSTEMS

Energy available in different forms from diverse natural sources can be converted into electrical energy using proper arrangement as described above. During this conversion process, some energy is always lost since it is converted to a different form than electrical energy. Consequently, electrical energy output is always lower than energy input. An electric power generation system's efficiency is the ratio of generated output electrical energy and input energy.

$$\text{Efficiency, } \eta = \frac{\text{Output electrical energy}}{\text{Input energy}} \times 100\,[\%] \tag{1.1}$$

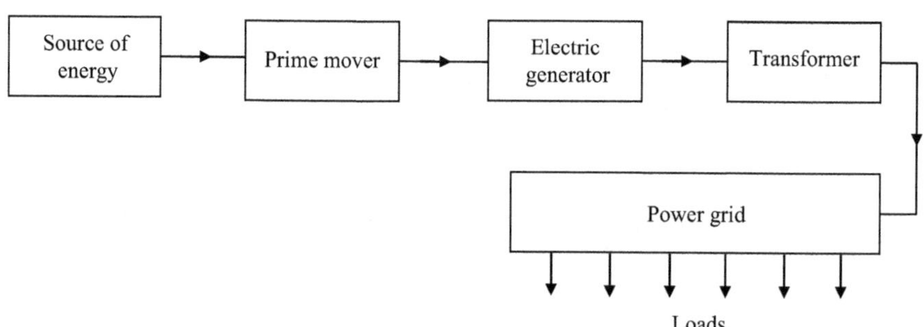

FIGURE 1.1 Simplified arrangement used for the generation of electrical energy from different energy sources.

The efficiency of an electrical power generation system can also be determined from the ratio of output electrical power and the input power, since power is simply the rate of flow of energy.

$$\text{Efficiency}, \eta = \frac{\text{Output electrical power}}{\text{Input power}} \times 100\,[\%] \tag{1.2}$$

All electrical power generation systems have efficiency less than 100% due to loss of some input energy during the conversion process.

Example 1.1

A three-phase synchronous generator delivers 800 [kW] power to a three-phase load. Mechanical power supplied to the synchronous generator is 900 [kW]. Determine the efficiency of the synchronous generator and energy lost per hour.

Solution

Input mechanical power = 900 [kW]

Output electrical power = 800 [kW]

$$\text{Efficiency}, \eta = \frac{\text{Output electrical power}}{\text{Input power}} \times 100 = \frac{800}{900} \times 100 = 88.89[\%]$$

Power lost = Energy lost per second = 100 [kW] = 100×10^3 [W]

Energy lost per hour = Energy lost per second $\times$ 3600 = $100 \times 10^3 \times 3600$

= 360×10^6 [J]

= 360 [MJ]

1.6 WORLD ELECTRICAL ENERGY GENERATION FROM DIFFERENT ENERGY SOURCES

Figure 1.2 shows percentages of electric energy generated from different energy sources in 2021. It can be observed that more than 70% of the electricity generated in the world is from fossil fuels and nuclear sources. Coal accounts for around 36% of electricity production globally, making it the most important means of power generation in the world. Nuclear fuels such as uranium account for around 10% of world electricity production.

World electric energy generation from renewable sources such as hydroelectric, geo-thermal, solar, tide, wind, and wave sources in 2021 was less than 30% of the total electrical energy generated, and this share will increase in the future as renewable electric energy sources are potentially more environmentally friendly and more abundant than non-renewable sources such as coal and natural gas.

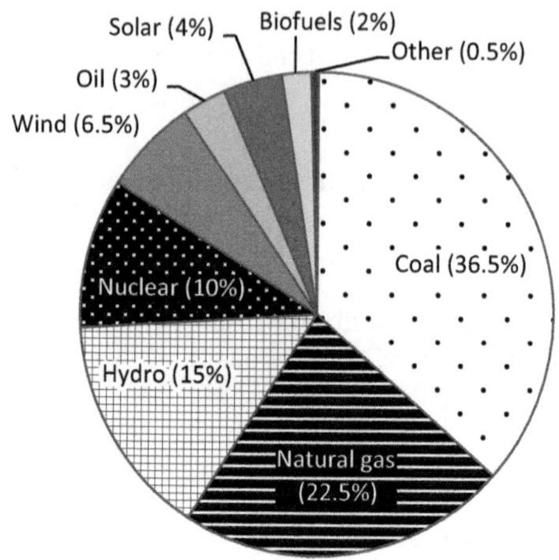

FIGURE 1.2 World electrical energy generation from different energy sources in 2021.

1.7 SUMMARY

The most useful form of energy among the different forms of energy available in nature is the electrical energy. It supports the modern technological civilization and helps in its advancement. Instantaneous availability, easy transmittability, easy controllability, easy convertibility, and pollution-free nature at the place of utilization are some of the characteristics which make electrical energy extremely beneficial. The consumption of electrical energy has been rising steadily, and electrical energy will become increasingly important in the future. It is therefore extremely useful for non-electrical engineers to have improved understanding of basic characteristics of electrical power systems. The purpose of this reference is to provide that knowledge.

REVIEW QUESTIONS

1.1 What are the advantages of electrical energy over other forms of energy?
1.2 What are the different phases in the development of human civilization?
1.3 Describe how the harnessing of various sources of energy has led to the development of human civilization.
1.4 Describe the general arrangement used for the conversion of other sources of energy into electrical energy in power plants.
1.5 Which are the principal sources of energy used for electrical energy generation?
1.6 Compare the different energy sources used for the generation of electrical energy.
1.7 Describe what is meant by efficiency of electrical energy generation systems.

1.8 Why is the efficiency of any electrical energy generation system always less than 100%?

1.9 What are the problems associated with the use of fossil fuel sources for the generation of electricity?

1.10 Why is the percentage of electricity generated from renewable energy sources increasing with time?

2 Structure of Electric Power Systems

2.1 INTRODUCTION

An electric power system transforms other forms of energy into electrical energy and transmits the electrical energy generated to the consumers for utilization. The production and transmission of electricity is relatively efficient and inexpensive. Electricity is not easily stored unlike other forms of energy, and it must generally be used as it is being produced. When we turn on a lighting system, the electricity it utilizes was generated the moment before.

Any electric power system consists of the following main sections:

1. Generation,
2. Transmission and sub-transmission, and
3. Distribution:
   - Primary distribution, and
   - Secondary distribution.

Simplified representation of the structure of an electric power system is shown in Figure 2.1. Generation is the production of electricity at the power plant. Transmission comprises of a set of transformers to raise the generated voltage to higher levels at the transmission substation and a system of long-distance power transmission lines. Transmission voltage levels are reduced to sub-transmission voltage levels at the high-voltage substation during the next stage of journey of electricity toward the load. Close to the load center, the sub-transmission voltage is stepped down to a lower primary distribution voltage level at the distribution substation. The primary distribution lines then carry the power to consumer premises, where the transformer drums lower the primary distribution voltage level to the secondary distribution voltage level or the utilization voltage level for the use by consumer's equipment.

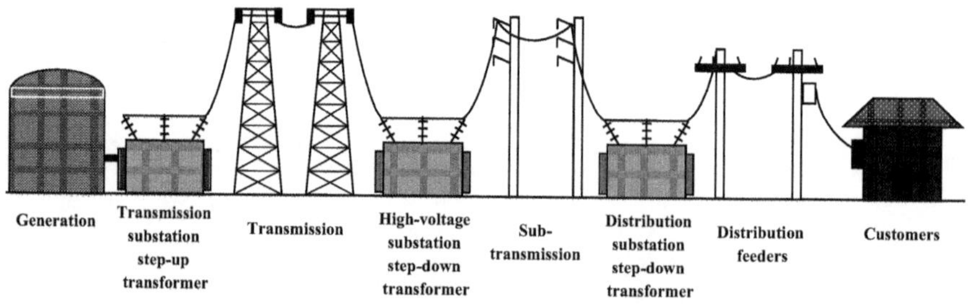

FIGURE 2.1 Simplified structure of an electric power system [31].

DOI: 10.1201/9781003432340-2

Power plants are sometimes located away from heavily populated areas and close to the source of fuel they use. The remote location close to the fuel source generally results in significant cost savings. Transmission of electricity long distances is carried out at much higher voltage levels than is practical for the utilization by most consumers in order to minimize power loss in the transmission lines.

The following sections describe each main portion of the power system mentioned above in more detail.

2.2 GENERATION OF ELECTRICAL ENERGY

Generation is the process of conversion of energy occurring in various forms in nature into electrical energy in the power plants. The following are the main types of power plants:

- Thermal power plants,
- Hydroelectric power plants,
- Gas turbine power plants,
- Nuclear power plants.

Thermal power plants and nuclear power plants use chemical energy of the fossil fuels, such as coal and nuclear energy of radioactive substance, such as uranium, respectively, for the generation of steam at high pressure and temperature from water, which is then used for the production of electrical energy by means of steam turbine connected to the three-phase synchronous generator. Hydroelectric power plants convert energy of moving water into electrical energy by means of hydraulic turbine connected to the three-phase synchronous generator. Gas turbine power plants generate electrical energy by burning fuel natural gas (methane) in compressed air and then passing the combustion products at high pressure and temperature through a gas turbine connected to the three-phase synchronous generator. The working of each of these types of power plants is described below.

2.2.1 Thermal Power Plants

Simplified working of a thermal power plant is shown in Figure 2.2. A thermal power plant converts heat energy of combustion of fossil fuel, such as coal, into electrical energy. The heat generated during combustion of pulverized coal is utilized in the boiler for the production of steam from water. The steam produced is at high pressure, and it is expanded in steam turbine, which produces mechanical power. The steam turbine drives the three-phase synchronous generator connected to it, which converts mechanical energy into electrical energy in the three-phase form. The exhaust steam after expansion in steam turbine is condensed in a condenser, and the water produced from steam condensation is fed back into the boiler. The gaseous products of combustion in boiler furnace are carried to the chimney and then to outside air through a dust collector and other cleansing apparatus.

Thermal power plants are generally located in places where there is an abundant supply of coal for steam production and water for cooling purpose. These plants are

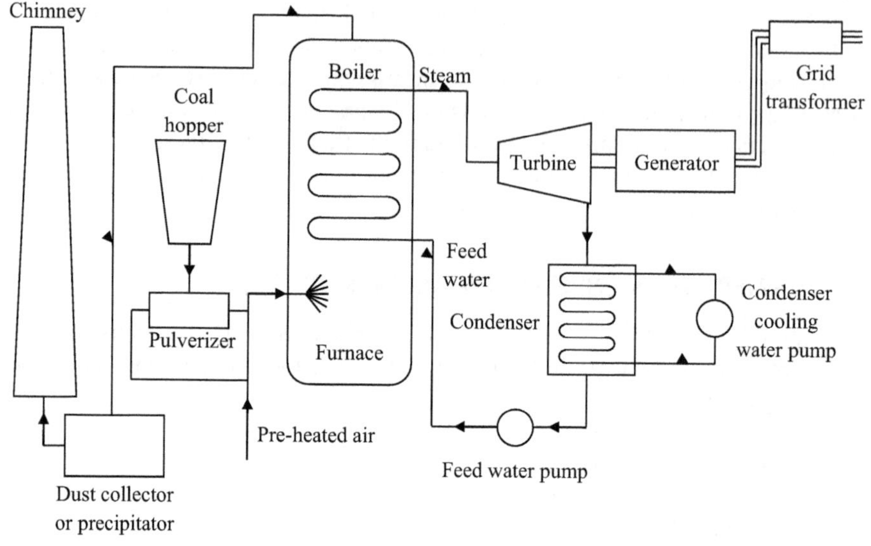

FIGURE 2.2 Simplified working of a thermal power plant [XVI].

generally constructed to have large power generation capacity, and power generation capacities of 1000 [MW] and above are common. The largest thermal power plant in the USA has installed capacity more than 3500 [MW].

The main advantages of thermal power plants are the use of comparatively cheaper fuel, such as coal, lower initial construction costs, lesser space requirement than hydroelectric power plants of similar capacity, and possibility of construction near load centers to reduce transmission losses with fuel transported to their place of location. The disadvantages of thermal power plants are higher maintenance and operating costs in comparison to hydroelectric and nuclear power plants, requirement of large quantity of water for steam generation in the boiler and steam condensation in the condenser, requirement of handling large amount of coal, problems in disposal of ash generated from coal combustion, and atmospheric pollution from smoke generated during coal combustion and from waste heat produced during the operation of the plant [23,33].

Example 2.1

A thermal power plant has an overall efficiency of 30%. It consumes 12500 [T] of coal each year for power generation. The coal has a heating value of 24000 [kJ/kg] (= 5742 [kcal/kg]). Determine the average power generated assuming that the plant runs throughout the year.

Solution

$$\text{Coal consumed per year} = 12500\,[\text{T}] = 12500 \times 10^3\,[\text{kg}]$$

$$\text{Heating value of coal} = 24000\ [\text{kJ / kg}]$$

$$\text{Total heat produced from coal combustion} = 24000 \times 12500 \times 10^3$$

$$= 300 \times 10^9\ [\text{kJ}]$$

$$\text{Coal heat converted to electrical energy} = 0.3 \times 300 \times 10^9 = 90 \times 10^9\ [\text{kJ}]$$

We have:

$$1\ [\text{kWh}] = 3.6 \times 10^6\ [\text{J}] = 3.6 \times 10^3\ [\text{kJ}]$$

$$\text{Total electrical energy generated in kWh} = \frac{90 \times 10^9}{3.6 \times 10^3} = 25 \times 10^6\ [\text{kWh}]$$

$$\text{Average power generated in kW} = \frac{\text{Total electrical energy generated in kWh}}{\text{Hours per year}}$$

$$= \frac{25 \times 10^6}{24 \times 365} = 2854\ [\text{kW}]$$

$$\text{Average power generated in MW} = \frac{2854}{10^3} = 2.854\ [\text{MW}]$$

2.2.2 Hydroelectric Power Plants

Hydroelectric power plants are popular since the non-renewable energy resources, such as coal and oil, are being depleted with the passage of time. Simplified working of a hydroelectric power plant is shown in Figure 2.3.

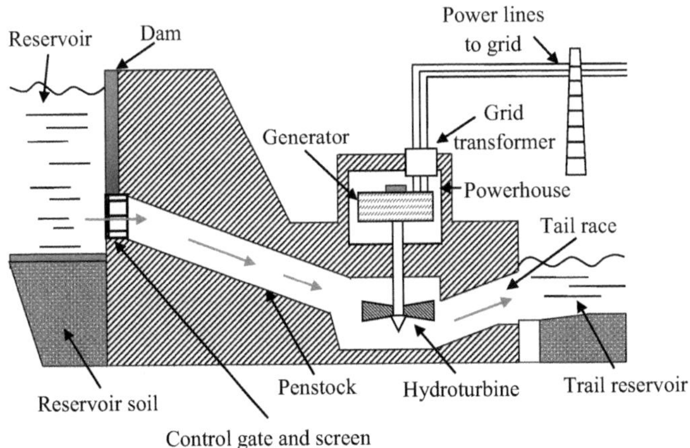

FIGURE 2.3 Simplified working of hydroelectric power generation [I].

Hydroelectric power plants harness potential energy of water at a high level for the generation of electrical energy. These plants are generally located in hilly areas where large reservoirs of water at a high head can be obtained by constructing a dam across the river. The reservoir acts as a source of water during the dry season for electricity generation, and it stores water for future use during the rainy season. Thus, reservoirs permit hydroelectric power plant operation at its rated capacity throughout the year [33,50]. Spillways are provided adjacent to the dams to allow excess water in reservoirs to be discharged into the downstream river without being used in electrical energy generation.

Conduits carry the water from the dam to large steel pipe structure called penstock, which brings water to the hydroturbine. The hydroturbine captures the kinetic energy of the falling water and converts it into mechanical energy at the turbine shaft. The hydroturbine drives the three-phase synchronous generator coupled to it, which converts mechanical energy of the turbine into electrical energy in the three-phase form [4,12,33]. After passing through the hydroturbine, water is discharged into the tailrace, which channels it into the downstream river.

The power generated in a hydroelectric power plant is determined from the following equation:

$$P = \rho g Q H \ [\text{W}]$$

$$= 9810 QH \ [\text{W}] \tag{2.1}$$

where:

Q = Water flow rate through the turbine (m^3/s),

ρ = Density of water (1000 [kg/m^3]),

g = Acceleration due to gravity (9.81 [m/s^2]),

H = Elevation difference between upper reservoir and lower trail reservoir water levels (m).

The main advantages of hydroelectric power plants are more useful working life, lower operating costs due to no fuel requirement, no problem of coal handling and ash disposal, clean and pollution-free operation, robust and simpler construction, higher reliability compared to thermal power plants, higher efficiency, no reduction in efficiency with age, very low start-up time of the order of a few minutes, no standby losses, faster generation control, and lower maintenance costs. The disadvantages of hydroelectric power plants are considerably higher per kW capacity construction costs in comparison to thermal power plants, longer construction time, lower capacity in comparison to thermal power plants, longer transmission line requirements and higher transmission losses due to remote location near supply of water in hilly areas, dependence on weather conditions for availability of water, and much larger area requirement due to reservoir construction.

Example 2.2

The water flow rate through a hydroelectric power plant turbine is approximately 100 [m³/s], and the elevation difference between upper reservoir and lower trail reservoir is 80 [m]. Determine the electrical power generated assuming plant efficiency to be 75%.

Solution

Water flow rate through the turbine:

$$Q = 100 \left[\mathrm{m}^3/\mathrm{s}\right]$$

Elevation difference between upper reservoir and lower trail reservoir:

$$H = 80\ [\mathrm{m}]$$

Plant efficiency:

$$\eta = 75\%$$

$$\text{Power generated} = P = \eta \times 9810 \times Q \times H = 0.75 \times 9810 \times 100 \times 80$$

$$= 58860000\ [\mathrm{W}] = 58.86\ [\mathrm{MW}]$$

2.2.3 Gas Turbine Power Plants

In a gas turbine power plant, electrical energy is generated by burning the fuel natural gas, such as methane in compressed air and then passing the combustion products through a gas turbine connected to the three-phase synchronous generator. Simplified working of a gas turbine power plant is shown in Figure 2.4.

In the gas turbine power plant, atmospheric air is first compressed in a compressor. The compressed air is then used to burn natural gas fuel in a combustion chamber. The high-temperature and high-pressure combustion products are passed through a gas turbine to do the mechanical work of rotating the turbine. The gas turbine drives the three-phase synchronous generator coupled to it, which converts mechanical energy of the turbine into three-phase electrical power. The gaseous combustion products are released into the atmosphere after doing mechanical work in the turbine. The mechanical energy produced in the turbine is partly utilized to compress the air.

The main advantages of gas turbine power plants are the use of cheaper natural gas fuel, faster installation, smaller size in comparison to a steam turbine power plant, absence of a boiler, quick starting and shut down time, smooth running, load following ability, and possibility of using a range of liquid and gaseous fuels including synthetic fuels and higher power generation efficiency from the possibility of using gas turbine exhaust for steam generation in boiler for driving another steam turbine connected to the three-phase synchronous generator. The disadvantages of gas turbine power plants are high operating temperatures imposing severe restriction

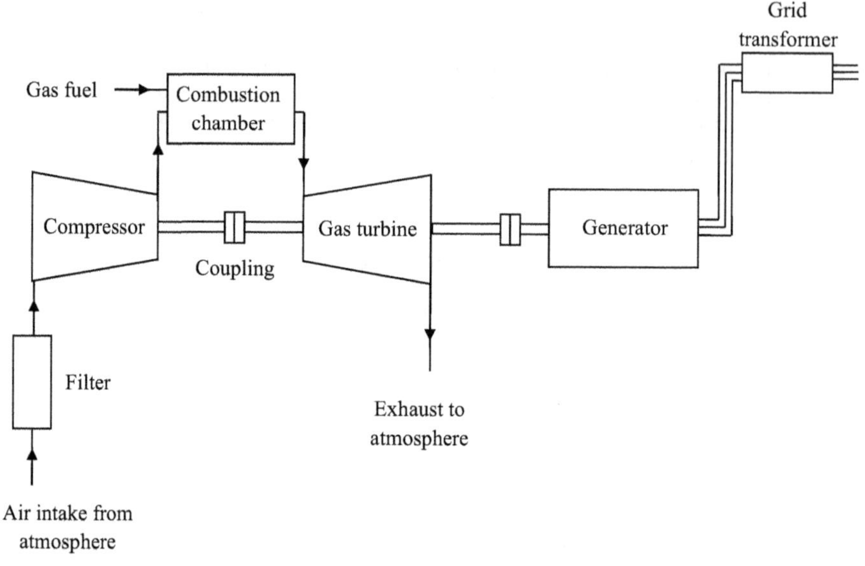

FIGURE 2.4 Simplified working of gas turbine power plant [XVII].

on the servicing conditions of the plant, requirement of special cooling methods for blades of the turbine due to the severity of operating temperatures and pressures, high noise levels, and incapability of using solid fuels, such as coal.

2.2.4 Nuclear Power Plants

Simplified working of pressurized water reactor (PWR) is shown in Figure 2.5. In such a reactor, steam is produced from water using heat energy generated during controlled nuclear fission in the reactor core. The primary loop of the nuclear reactor uses a pressurizer to prevent the formation of steam, which exerts high pressure to water to prevent it from boiling. In heat exchanger or steam generator, water in the primary loop that is under high pressure and temperature transfers that heat to water in the secondary loop. The high-pressure steam produced in the secondary loop is expanded in a steam turbine to provide mechanical power, which then powers the three-phase synchronous generator connected to it and generates three-phase electric power. After expansion, the exhaust steam is condensed in a condenser, and the condensed water is returned to the secondary loop for production of steam in the steam generator.

Almost 0.1% of the mass of the uranium-235 isotope is transformed into heat energy during the nuclear fission process. The energy produced during nuclear fission of one kilogram of uranium-235 isotope is the same as that produced from burning about 3000 [T] of coal. A fleet of railway cars is required to provide continuous supply of coal for the working of thermal power plant, whereas a few trucks can supply the annual fuel requirements of a nuclear power plant of a similar capacity. Around once a year, the nuclear reactor is shut down to replenish the uranium fuel consumed.

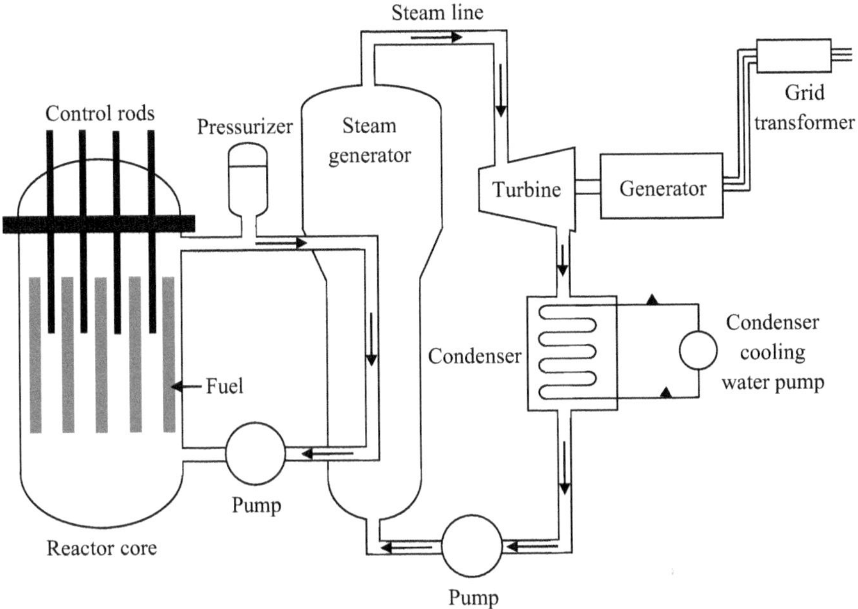

FIGURE 2.5 Simplified working of pressurized water nuclear reactor [27].

Simplified working of another type of nuclear reactor, called boiling water reactor (BWR), is shown in Figure 2.6. In a BWR, water is directly converted into steam in the reactor core, which is then expanded in steam turbine connected to a three-phase synchronous generator for the generation of three-phase electrical power. The pumping of coolant water through the reactor core continuously removes the heat produced during the nuclear fission reaction in the core of radioactive material. The coolant water boiling under high pressure produces steam. In such a nuclear reactor, heat exchanger is not required since the steam produced is circulated directly through the turbine. After expansion, the exhaust steam is condensed in a condenser, and the condensed water is pumped back into the reactor core for heat removal and generation of steam [48,49].

The main advantages of nuclear power plants are small amount of fuel requirement, no problem of continuous fuel transportation and storage as in thermal power plants, considerable savings in fuel transportation cost, lesser plant space requirement in comparison to other plant types of similar capacity, low operating costs, reduced transmission and primary distribution costs due to the possibility of location near load centers, higher operational reliability in comparison to thermal power plants, neat and quiet plant operation due to low fuel requirement, and availability of large deposits of nuclear fuel all over the world ensuring continued supply of electrical energy for millions of years in future. The disadvantages of nuclear power plants are higher initial capital costs, greater technical skill requirement in construction and commissioning of the plant, difficulty in disposal of radioactive by products of nuclear reaction, unsuitable operation in case of varying load conditions, more

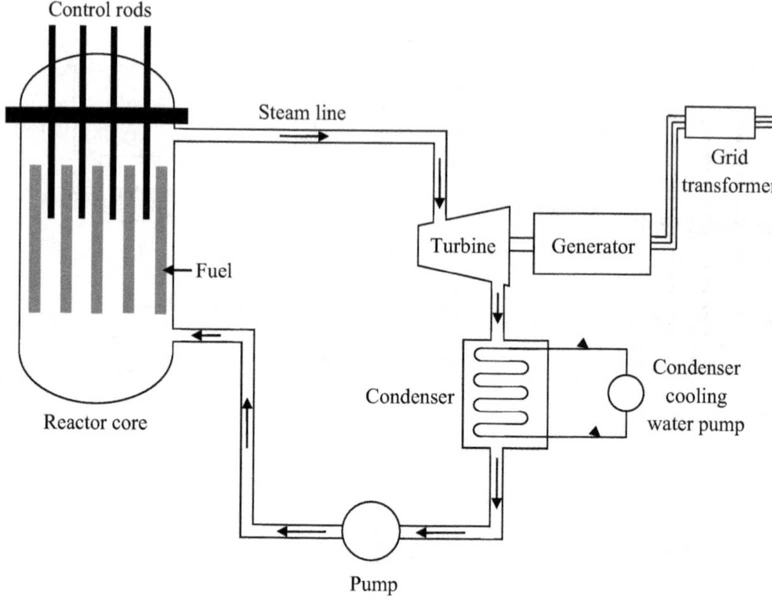

FIGURE 2.6 Simplified working of boiling water reactor [27].

cooling water requirement in comparison to thermal power plants of similar capacity, use of expensive fuel, and higher maintenance costs.

Example 2.3

A nuclear power plant generates 200 [MW]. The nuclear fission of each U-235 atom generates 200 [MeV] energy where $1\,[\text{eV}] = 1.6\times10^{-19}$ [J]. Determine the mass of uranium fuel consumed per hour assuming that the uranium fuel contains only fissile uranium atoms.

Solution

$$\text{Power generated in nuclear power plant} = 200\,[\text{MW}]$$

$$\text{Electrical energy generated per hour} = 200\times10^{6}\times3600 = 72\times10^{10}\,[\text{J}]$$

$$\text{Electrical energy from single fission} = 200\times10^{6}\times1.6\times10^{-19}$$

$$= 32\times10^{-12}\,[\text{J}]$$

Number of uranium atoms fissioned per hour

$$= \frac{\text{Electrical energy generated per hour}}{\text{Electrical energy from single fission}} = \frac{72\times10^{10}}{32\times10^{-12}} = 2.25\times10^{22}$$

Since 235 [g] uranium fuel contains 6.023×10^{23} atoms, we have:

$$\text{Mass of uranium fuel fissioned per hour} = \frac{235}{6.023\times10^{23}}\times2.25\times10^{22}$$

$$= 8.78\ [\text{g}]$$

2.3 TRANSMISSION AND SUB-TRANSMISSION

Power plants generating electrical energy are generally constructed away from load centers where consumers are located. The overhead power transmission system is used to transmit the electrical energy produced in power plants at various locations to the power distribution system via a sub-transmission system, which subsequently supplies electrical energy to consumer loads. Electricity generated in power plants generally has a voltage level less than 30 [kV] three-phase. Unit transformers raise the generation voltage level to transmission voltage level, such as 345 [kV] line-to-line or higher in the USA, in order to economically and effectively transmit large amounts of power to the load centers.

A grid is the network of these extremely high-voltage transmission lines in an area or state. Various neighboring grids in a region are connected using tie lines to form a regional grid, and various regional grids are also connected to form the national grid. The generated electrical power in power plants is directly fed into the local grid through a transformer. The individual grids operate independently. However, during conditions such as abrupt loss of generation or increase in load, power can be transferred from one grid to other using tie lines.

The transmission voltage level is subsequently stepped down to the sub-transmission voltage level at high-voltage substations. The sub-transmission network operates at voltage levels less than transmission level voltage, for example, between 69 and 138 [kV] line-to-line in the USA. The sub-transmission network then carries power to the distribution substation located near the load center, where the voltage level is further reduced for primary distribution [15,31,48]. Very large industrial customers receive power directly from the transmission system, whereas large industrial customers receive power from the sub-transmission network.

Figure 2.7 shows a simplified representation of transmission tower supporting the three-phase transmission lines. The ground wires are located and attached at the top of the transmission tower. They protect the three-phase transmission lines from lightning strikes and consequent damage and power surges.

2.4 DISTRIBUTION

The distribution system links the distribution substation to the consumer feeders to supply power to domestic loads. The primary distribution voltage levels in the USA are between 4 and 34.5 [kV] line-to-line. Some small industrial customers receive power directly from the primary distribution system at primary distribution voltage levels. The primary distribution voltage is further reduced to secondary distribution or utilization voltage level using a distribution transformer close to the consumer's

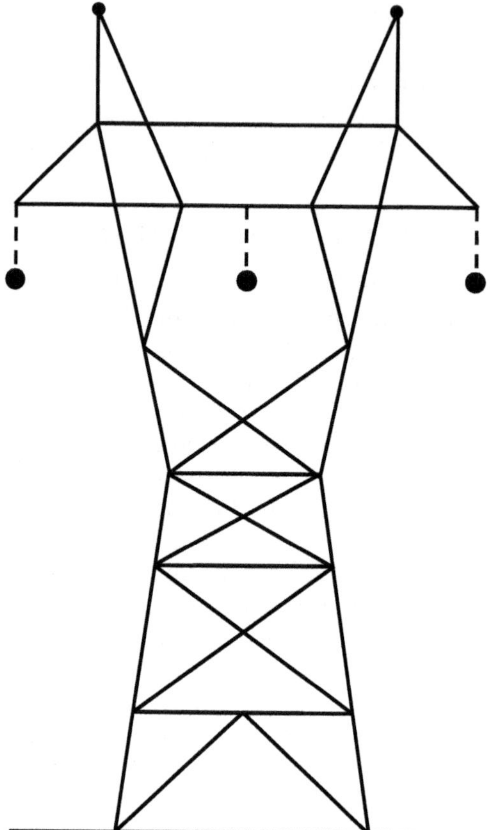

FIGURE 2.7 Simplified representation of a transmission line tower supporting the three-phase power transmission lines [18].

premises. The secondary distribution voltage level at which power is supplied to domestic loads is mainly 120 [V] single-phase in the USA.

In order to have a balanced power system operation, loads are balanced in the three phases of the distribution network. For example, 100 houses each are connected to each of the three phases out of the 300 houses in a locality in order to balance the loads in the three phases of the power distribution system. At higher voltage levels, the balancing effect is more clearly apparent. The total load in each of the three phases of a large town area is for all practical purposes balanced.

Distribution systems can be either overhead type or underground type. Many newly constructed residential areas have an underground power distribution system due to aesthetic reasons.

2.5 UTILIZATION

The main load types for the utilization consist of industrial, commercial, and residential loads. Industrial loads mainly consist of loads of large industries, such as metals, chemicals, paper products, and food processing. Induction motors form a large part

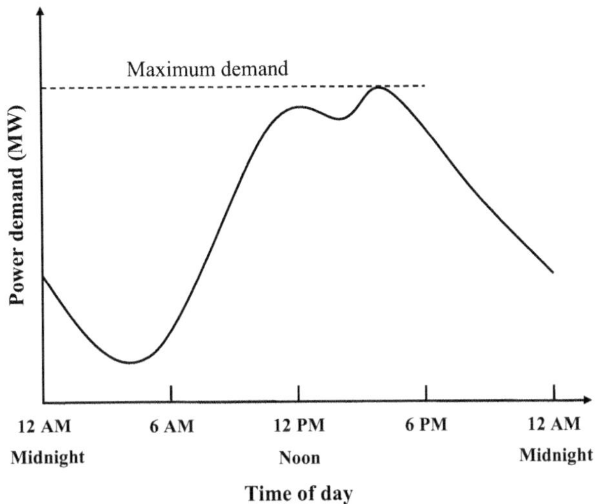

FIGURE 2.8 Daily load demand curve of a power system [38].

of industrial loads. Commercial loads mainly consist of lighting, heating, and cooling loads of offices, shops, and schools. House equipment, such as lighting systems, refrigerators, microwaves, induction stoves, water heaters, space heating systems, and air conditioners, make up the large part of residential loads.

Depending on the requirements of their activities, consumers consume small or large amounts of electrical power, and each consumer has a particular load demand requirement. Consumer demands are, therefore, unpredictable, and a power system's load varies over time. The generation capacity of power plants should be such that it can supply the maximum possible demand requirement of the consumers [38,48]. The typical daily load demand variation with time of a power system is shown in Figure 2.8.

2.6 HISTORY OF ELECTRIC POWER SYSTEMS

The first electric power system was developed by Thomas Edison, which started working in the year 1882 at New York City's Pearl Street Station. The system provided direct current (DC) electricity produced at 110 [V] using DC generators powered by steam engines for lighting a one square mile area having a total power demand of 30 [kW]. The development of the DC motor resulted in further development of a DC electric power system. Voltage drops in the transmission system, however, increased and became more intolerable as a consequence of an increase in load and the higher current that occurred in low-voltage DC systems. It also became increasingly difficult to locate new generating stations close to the loads being supplied due to limited space availability in the city area. A possible solution to the problem was generating power at a distant location, transmitting it at a higher voltage level to reduce transmission losses, and then lowering the voltage level close to the loads for the utilization purpose. The lack of a voltage level transformation device for DC power systems, however, rendered this solution impractical.

The experimental alternating current (AC) transmission system was developed by L. Gauland and J. D. Gibbs in 1881, which employed transformers to increase the generated AC voltage level for transmission and then decrease it close to the place of utilization for lower transmission losses, thus making long-distance electric power transmission practicable. The first experimental AC distribution system in the USA was constructed in Great Barrington, Massachusetts, in 1886 to power 150 lamps for lighting purposes. The first single-phase AC transmission system in the USA was constructed between Oregon City and Portland in 1889. It was 13 miles long with a transmission voltage of 4 [kV], and the transmitted power was utilized only for the lighting purpose.

The polyphase AC systems soon became popular due to Nikola Tesla's invention of polyphase induction and synchronous motors in 1888, the lack of commutation in AC generators making generation of more power at a higher voltage level possible, the availability of transformers to transform the voltage level for lower transmission losses in long-distance power transmission, and the constancy of supplied power in polyphase systems, such as the three-phase system. The first three-phase transmission system in the United States was constructed in Southern California in 1893, which had a length of 7.5 [miles] and a transmission voltage of 2.3 [kV].

The advantages of interconnecting the isolated power systems to supply the peak load demand from combined generation for greater economy and improved reliability became apparent with the expansion of isolated electric utilities generating power in different regions. In the early years of twentieth century, power was generated at a wide range of frequencies, such as 25, 50, 60, 125, and 133 [Hz]. It was required to have power generation at a single frequency for the interconnection of several isolated power systems. In the USA and Canada, a frequency of 60 [Hz] was accepted as the standard for the generation of AC power. Europe, on the other hand, accepted 50 [Hz] as the standard frequency for AC power generation. In developed nations, such as the USA, regional power systems are interconnected to create a national power grid system, and power transfer is possible between various regions during normal operations and in times of emergency for improved economy, reliability, and security.

Due to the ever-rising demand for power, it was required to have power transmission at higher voltage levels. The maximum three-phase AC transmission voltage in the USA increased from 2.3 [kV] in 1893 to 40 [kV] in 1900, 150 [kV] in 1920, 345 [kV] in 1960, and 765 [kV] in 1970. Research for the transmission of power at higher voltage levels between 1000 and 1500 [KV] is in progress. Furthermore, the power industry also adopted a few voltage levels as the standard in order to avoid using an unlimited number of voltage levels for transmission, distribution, and utilization.

For long-distance AC power transmission, it has been found to be more cost-effective to convert AC power to DC, transmit electricity through DC transmission lines, and then invert DC electricity back to AC at the other end of the transmission line. DC transmission lines are more cost-effective for transmission distances greater than 500 [km]. DC transmission lines have no reactance, and they permit transmission of more power than AC transmission lines over the same size of the conductor. The first high-voltage DC transmission system was installed in Sweden in 1954 between Swedish Island of Gotland and Swedish mainland, and it was 60 miles long with a transmission voltage of 100 [kV]. The first high-voltage DC transmission system in the USA was constructed between Oregon and California in 1970, which was 850 [miles] long having a transmission voltage of 800 [kV]. Numerous installations of high-voltage DC transmission lines have been constructed worldwide since then and are in use.

From its simple beginning at the end of the nineteenth century, the electric power system has developed into one of the largest industries in the world and is developing further.

2.7 SUMMARY

The main sections of an electric power system include generation, transmission, primary distribution, and secondary distribution or utilization. Generation is the production of electrical energy in the power plants. Transmission lines connect various power plants to the distribution systems, which subsequently supply electricity to various loads for the utilization.

The main types of power plants are thermal power plants and gas turbine power plants converting chemical energy of fuel into electrical energy, hydroelectric power plants converting hydraulic energy of water into electrical energy, and nuclear power plants converting nuclear energy of radioactive substance into electrical energy.

The electrical power generated in power plants generally has a voltage level less than 30 [kV]. For the economical and effective transfer of power from generating stations to the load centers, the generated voltage levels are raised to much higher transmission voltage levels using unit transformers. A regional power system in countries, such as the USA, is connected to power systems of neighboring regions at the transmission voltage level using tie lines, which allows the transfer of power between different regional power systems mainly during emergencies. The transmission voltage level is subsequently stepped down to the sub-transmission voltage level at high-voltage substations. The transformers in distribution substations located close to load centers reduce the sub-transmission level voltage to the primary distribution level. Near customer premises, the primary distribution voltage level is reduced to the secondary distribution or utilization voltage level to supply the loads.

Due to uncertain demands of consumers, the overall load on the electrical power system varies over time. The power generation capacity of power plants should be capable of meeting the highest possible load demand of the consumers.

REVIEW QUESTIONS

2.1 What are the main parts of an electric power system?

2.2 What are the main types of power plants used for the bulk generation of electrical energy?

2.3 Describe the process of conversion of energy of fossil fuels into electrical energy in thermal power plants.

2.4 What is the use of a condenser in a thermal power plant?

2.5 Describe the main components and working of a hydroelectric power plant?
2.6 Describe the main advantages and disadvantages of hydroelectric power plants.
2.7 Describe the working of a gas turbine power plant.
2.8 Describe the process of conversion of nuclear energy into electrical energy in a pressurized water nuclear reactor.
2.9 Describe the process of conversion of nuclear energy into electrical energy in a boiling water nuclear reactor.
2.10 Write a short note on the transmission of electrical power.
2.11 Write a short note on the distribution of electrical power.
2.12 Write a short note on the utilization of electrical power.

3 AC Circuit Fundamentals
Single-Phase Circuits

3.1 INTRODUCTION

Electric current can flow in an electrical circuit in two different ways: alternating current (AC) or direct current (DC). AC is the current which periodically reverses direction, unlike direct current, which only flows in one direction. The kind of electric current which is generated by power plants is AC. AC has the main benefit of being relatively simple to transform power system voltage from one level to another using a transformer. It permits power transmission at very high voltages before being reduced to safer voltages for commercial and domestic use. This results in the minimization of energy losses in power transmission and distribution.

The fundamentals of single-phase AC circuits are described in the present chapter, whereas the fundamentals of the three-phase AC circuits are described in the next chapter.

3.2 VOLTAGE, CURRENT, AND POWER IN SINGLE-PHASE AC SYSTEMS

A single-phase AC voltage source supplying power to an electrical load is shown in Figure 3.1.

The following equations represent the instantaneous voltage and current supplied by the AC source:

$$v = V_{\max} \sin \omega t$$

$$i = I_{\max} \sin\left(\omega t - \phi\right)$$

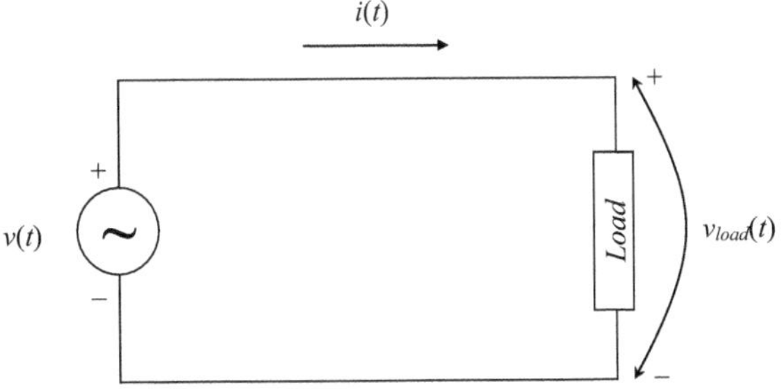

FIGURE 3.1 Simple single-phase AC circuit.

DOI: 10.1201/9781003432340-3

where

v = instantaneous value of sinusoidal voltage at time t in volt [V]
V_{max} = peak value or amplitude of source voltage in volt [V]
i = instantaneous value of sinusoidal current at time t in ampere [A]
I_{max} = peak value or amplitude of current supplied by the source in ampere [A]
f = frequency of AC voltage source in s^{-1}
$\omega = 2\pi f$ = angular frequency of voltage source in rad/s
ϕ = phase angle by which v waveform leads i waveform with ϕ assumed positive when v leads i

In the sinusoidal single-phase AC circuit shown above, power supplied to the load is calculated as follows:

$$
\begin{aligned}
p = vi &= V_{max}\ I_{max}\ \sin\omega t\ \sin(\omega t - \phi) \\
&= \frac{V_{max}}{\sqrt{2}}\frac{I_{max}}{\sqrt{2}}\left[\cos\phi - \cos(2\omega t - \phi)\right] \\
&= VI\left[\cos\phi - \cos(2\omega t - \phi)\right] \\
&= VI\cos\phi - VI\cos(2\omega t - \phi)
\end{aligned} \tag{3.1}
$$

where

$V = \frac{V_{max}}{\sqrt{2}}$ = root-mean-square (RMS) value of the source voltage [V]
$I = \frac{I_{max}}{\sqrt{2}}$ = root-mean-square (RMS) value of the current supplied by the source [A]

ϕ in the above equation is constant. p has two components in the above equation: the first component $VI\cos\phi$ remains constant with time, whereas the second component $VI\cos(2\omega t - \phi)$ pulsates at an angular frequency of 2ω.

The above p equation is integrated for one cycle between 0 and $T = \frac{2\pi}{\omega} = \frac{1}{f}$ and then divided by T to find the average value of the time-varying power $p_{average}$:

$$
p_{average} = \frac{1}{T}\int_0^T p\ dt = VI\ \cos\phi \tag{3.2}
$$

Power supplied by the source pulsates around the average value $p_{average}$ obtained above. $p_{average}$ is always positive; however, instantaneous power p can become negative or zero during certain times in each cycle. The waveforms of v, i, and p are shown in Figure 3.2.

In the case of a resistive load, ϕ equals 0°, and $p_{average}$ is positive.
In the case of an inductive load, ϕ equals 90°, and $p_{average}$ is zero.
Similarly, in the case of a capacitive load, ϕ equals –90°, and $p_{average}$ is zero.

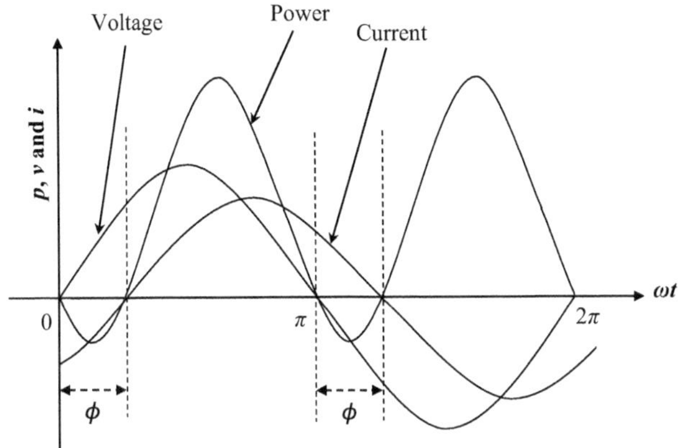

FIGURE 3.2 Waveforms of v, i, and p[14].

3.3 PHASOR REPRESENTATION

Instantaneous values of source voltage and current are expressed using RMS values as follows:

$$v = V_{\max} \sin \omega t = \sqrt{2} V \sin \omega t \tag{3.3}$$

$$i = I_{\max} \sin(\omega t - \phi) = \sqrt{2} I \sin(\omega t - \phi) \tag{3.4}$$

V and I in the above equations are RMS values of sinusoidal voltage and current as described earlier.

We next use Euler's formula to obtain a phasor representation of sinusoidal voltage and current waveforms, which states that:

$$e^{j(\omega t + \theta)} = \cos(\omega t + \theta) + j \sin(\omega t + \theta)$$

v and i can now be written as:

$$\begin{aligned} v &= \text{Im}\left[\sqrt{2} V e^{j\omega t}\right] = \text{Im}\left[\sqrt{2} \times \boldsymbol{V} e^{j\omega t}\right] \\ \boldsymbol{V} &= V = |\boldsymbol{V}| = V e^{j0^\circ} = V\angle 0^\circ = |\boldsymbol{V}|\angle 0^\circ \end{aligned} \tag{3.5}$$

$$\begin{aligned} i &= \text{Im}\left[\sqrt{2} I e^{j\omega t} e^{-j\phi}\right] = \text{Im}\left[\sqrt{2} \times \boldsymbol{I} e^{j\omega t}\right] \\ \boldsymbol{I} &= I e^{-j\phi} = I\angle -\phi = |\boldsymbol{I}|\angle -\phi \end{aligned} \tag{3.6}$$

The representations $\boldsymbol{V}$ and $\boldsymbol{I}$ are known as phasors. It is possible to construct sinusoidal waveforms of voltage and current from corresponding phasors by multiplication with $\sqrt{2}e^{j\omega t}$ as shown above. Phasors are represented by bold letters. Their magnitudes are represented by non-bold letters or bold letters enclosed within symbol | |.

In general, we can write:

$$\boldsymbol{V} = V\angle\ \theta_v = |\boldsymbol{V}|\angle\theta_v \tag{3.7}$$

$$\boldsymbol{I} = I\angle\theta_i = |\boldsymbol{I}|\angle\theta_i \tag{3.8}$$

where

$|\boldsymbol{V}|$ / V and $|\boldsymbol{I}|$ / I = magnitudes of phasors $\boldsymbol{V}$ and $\boldsymbol{I}$

θ_v and θ_i= phase angles of $\boldsymbol{V}$ and $\boldsymbol{I}$ in degrees

We assume that general waveforms of voltage and current are expressed as:

$$v = V_{\max}\sin(\omega t + \theta_v) = \sqrt{2}V\sin(\omega t + \theta_v) = \sqrt{2}|\boldsymbol{V}|\sin(\omega t + \theta_v) \tag{3.9}$$

$$i = I_{\max}\sin(\omega t + \theta_i) = \sqrt{2}I\sin(\omega t + \theta_i) = \sqrt{2}|\boldsymbol{I}|\sin(\omega t + \theta_i) \tag{3.10}$$

Using the approach described above, we can represent v and i as phasors $\boldsymbol{V}$ and $\boldsymbol{I}$ by the following expressions:

$$\boldsymbol{V} = V\angle\theta_v = |\boldsymbol{V}|\angle\theta_v \tag{3.11}$$

$$\boldsymbol{I} = I\angle\theta_i = |\boldsymbol{I}|\angle\theta_i \tag{3.12}$$

Example 3.1

Determine the corresponding voltage and current phasors for the following voltage and current waveforms:

$$v = \sqrt{2}(120)\sin(120\pi t + 30°)[\mathrm{V}]$$

$$i = \sqrt{2}(20)\sin(120\pi t - 60°)[\mathrm{A}]$$

Solution

The voltage waveform is:

$$v = \sqrt{2}(120)\sin(120\pi t + 30°)[\mathrm{V}]$$

The corresponding voltage phasor is:

$$\boldsymbol{V} = 120\angle 30°\ [\mathrm{V}]$$

The current waveform is:

$$i = \sqrt{2}(20)\sin(120\pi t - 60°)[\text{A}]$$

The corresponding current phasor is:

$$\boldsymbol{I} = 20\ \angle -60°\ [\text{A}]$$

3.4 REAL, REACTIVE, AND APPARENT POWER

The following expression was obtained earlier for instantaneous power in single-phase AC circuits:

$$p = VI\cos\phi - VI\cos(2\omega t - \phi) = |\boldsymbol{V}||\boldsymbol{I}|\cos\phi - |\boldsymbol{V}||\boldsymbol{I}|\cos(2\omega t - \phi) \qquad (3.1)$$

The second term in the above equation is now expanded as follows:

$$\begin{aligned} p &= |\boldsymbol{V}||\boldsymbol{I}|\cos\phi - |\boldsymbol{V}||\boldsymbol{I}|[\cos 2\omega t\ \cos\phi + \sin 2\omega t\ \sin\phi] \\ &= |\boldsymbol{V}||\boldsymbol{I}|\cos\phi[1 - \cos 2\omega t] - |\boldsymbol{V}||\boldsymbol{I}|\sin\phi[\sin 2\omega t] \qquad (3.13) \\ &= \underbrace{P[1 - \cos 2\omega t]}_{I} - \underbrace{Q[\sin 2\omega t]}_{II} \end{aligned}$$

where

$$P = |\boldsymbol{V}||\boldsymbol{I}|\cos\phi \qquad (3.14)$$

$$Q = |\boldsymbol{V}||\boldsymbol{I}|\sin\phi \qquad (3.15)$$

In the above p equation, the first component pulsates around the average value P in each cycle; however, it never becomes negative. In sinusoidal single-phase AC circuits, P, the average value of p, is used to represent the average power supplied by the source. P is called the real power of a sinusoidal single-phase AC circuit. Its magnitude depends on $\cos\phi$. The unit of real power P is watt [W], where $1\ \text{W} = 1\ \text{V} \times 1\ \text{A}$.

The second component is positive half of the time and negative half of the time in each cycle. It has an average value of 0, and its maximum value is Q. This term represents the power that flows first from the source to the load and then back from the load to the source. This power which is constantly bouncing between the source and the load is called the reactive power. Reactive power is the energy which is first stored in an inductor's magnetic field or a capacitor's electric field and then released.

The reactive power of a load is represented by the peak value Q of that power component, which flows back and forth in the line. It has an average value of 0, and consequently, it is not capable of doing any useful work. The reactive power of the load therefore is $Q = |\boldsymbol{V}||\boldsymbol{I}|\sin\phi$. Q is positive for inductive loads and negative for capacitive loads by convention, as the phase angle ϕ between voltage and current phasors is positive for inductive loads and negative for capacitive loads. The unit of reactive power Q is volt-ampere reactive [VAR], where $1\ \text{VAR} = 1\ \text{V} \times 1\ \text{A}$.

The unit of Q is given a different name of VAR even when P and Q have the identical physical dimension of watt [W] in order to distinguish between active or real power P and non-active or reactive power Q. Figure 3.3 shows the waveforms of real, reactive, and total powers in a single-phase AC circuit.

The apparent power S supplied to the load is the product of the voltage across the load and the current flowing through the load. This is the power that the source appears to supply to the load if the phase angle difference between the voltage and current phasors is ignored.

The apparent power of a load is thus given by:

$$S = |\boldsymbol{V}||\boldsymbol{I}| \tag{3.16}$$

Apparent power is extremely useful as a unit for the rating of electrical apparatus such as the transformer.

The unit of apparent power is volt-ampere [VA], where $1\ \text{VA} = 1\ \text{V} \times 1\ \text{A}$. In order to distinguish it from real power and reactive power, the unit of apparent power is given a different name.

We also have:

$$S = \sqrt{P^2 + Q^2} = \sqrt{|\boldsymbol{V}|^2 |\boldsymbol{I}|^2 \cos^2\phi + |\boldsymbol{V}|^2 |\boldsymbol{I}|^2 \sin^2\phi} = |\boldsymbol{V}||\boldsymbol{I}| \tag{3.17}$$

More practical units of P, Q, and S are:

$$1\ \text{MW} = 10^3\ \text{kW} = 10^6\ \text{W}$$

$$1\ \text{MVAR} = 10^3\ \text{kVAR} = 10^6\ \text{VAR}$$

$$1\ \text{MVA} = 10^3\ \text{kVA} = 10^6\ \text{VA}$$

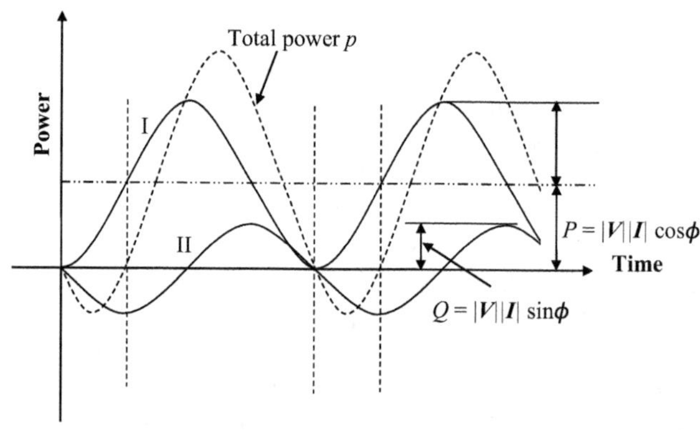

FIGURE 3.3 Waveforms of real, reactive, and total powers in a single-phase AC circuit [14].

3.5 COMPLEX POWER

Real power P and reactive power Q are sometimes represented together as complex power $\boldsymbol{S}$ for simplicity in computer calculations.

$$\boldsymbol{S} = P + jQ \tag{3.18}$$

The complex power $\boldsymbol{S}$ supplied to a load is calculated from the equation:

$$\boldsymbol{S} = \boldsymbol{V} \times \boldsymbol{I}^* \tag{3.19}$$

The asterisk in the above equation represents the complex conjugate operator.

We assume that voltage and current phasors are given by the following expressions:

$$\boldsymbol{V} = |\boldsymbol{V}| \angle \theta_v \tag{3.20}$$

$$\boldsymbol{I} = |\boldsymbol{I}| \angle \theta_i \tag{3.21}$$

The complex power $\boldsymbol{S}$ is then calculated as follows:

$$\begin{aligned} \boldsymbol{S} = \boldsymbol{V}\boldsymbol{I}^* &= |\boldsymbol{V}| \angle \theta_v \times |\boldsymbol{I}| \angle -\theta_i = |\boldsymbol{V}||\boldsymbol{I}| \angle \theta_v - \theta_i \\ &= |\boldsymbol{V}||\boldsymbol{I}| \cos(\theta_v - \theta_i) + j|\boldsymbol{V}||\boldsymbol{I}| \sin(\theta_v - \theta_i) \\ &= |\boldsymbol{V}||\boldsymbol{I}| \cos\theta + j|\boldsymbol{V}||\boldsymbol{I}| \sin\theta \\ &= P + jQ \end{aligned} \tag{3.22}$$

If the load impedance is given by phasor $\mathbf{Z} = Z\angle\theta = |\mathbf{Z}| \angle\theta$ as shown in Figure 3.4, we can use it to obtain an alternative expression for complex power.

From Ohm's law, we have for sinusoidal single-phase AC circuits:

$$\boldsymbol{V} = \boldsymbol{I}\,\boldsymbol{Z} \tag{3.23}$$

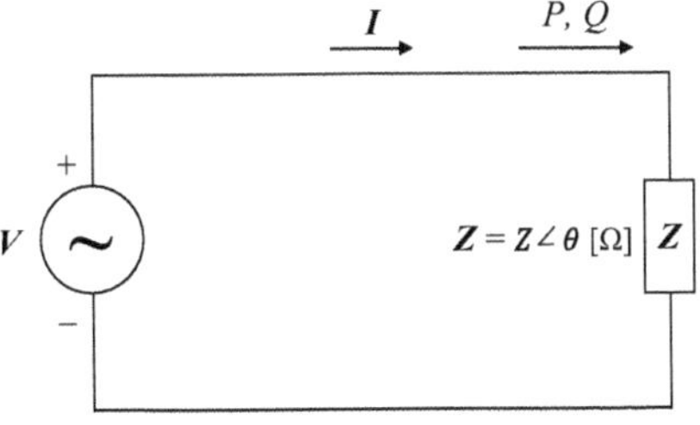

FIGURE 3.4 Phasor representation of load in a single-phase AC circuit.

where

$$\boldsymbol{Z} = |\boldsymbol{Z}| \angle\theta = \frac{1}{Y}\angle\theta = \frac{1}{|\boldsymbol{Y}|\angle-\theta} = \frac{1}{\boldsymbol{Y}} \tag{3.24}$$

$\boldsymbol{Y} = Y\angle-\theta = |\boldsymbol{Y}|\angle-\theta$ is the admittance of the load in phasor form, and it is the inverse of $\boldsymbol{Z}$. The impedances and admittances of resistors, inductors, and capacitors are shown in Table 3.1. ω is the angular frequency of the voltage source in rad/s as described above.

Substituting $\boldsymbol{V}$ obtained above in complex power expression, we obtain:

$$\boldsymbol{S} = \boldsymbol{V} \times \boldsymbol{I}^* = \boldsymbol{I}\,\boldsymbol{Z} \times \boldsymbol{I}^* = \boldsymbol{I}\boldsymbol{I}^*\boldsymbol{Z} = |\boldsymbol{I}|^2\,\boldsymbol{Z} \tag{3.25}$$

The impedance of a load can be expressed as:

$$\boldsymbol{Z} = R + jX = |\boldsymbol{Z}|\cos\theta + j|\boldsymbol{Z}|\sin\theta \tag{3.26}$$

We thus have:

$$\boldsymbol{S} = |\boldsymbol{I}|^2|\boldsymbol{Z}|\cos\theta + j|\boldsymbol{I}|^2|\boldsymbol{Z}|\sin\theta = P + jQ \tag{3.27}$$

The real and reactive powers of a load can also be expressed as:

$$P = |\boldsymbol{I}|^2|\boldsymbol{Z}|\cos\theta = |\boldsymbol{I}|^2\,R \tag{3.28}$$

$$Q = |\boldsymbol{I}|^2|\boldsymbol{Z}|\sin\theta = |\boldsymbol{I}|^2\,X \tag{3.29}$$

where

$R = |\boldsymbol{Z}|\cos\theta = Z\cos\theta =$ resistance of the load
$X = |\boldsymbol{Z}|\sin\theta = Z\sin\theta =$ reactance of the load

TABLE 3.1
Impedances and Admittances of Resistors, Inductors, and Capacitors

Circuit Element	Impedance	Admittance
Resistance R	$\boldsymbol{Z} = R$	$\boldsymbol{Y} = \frac{1}{R}$
Inductance L	$\boldsymbol{Z} = j\omega L$	$\boldsymbol{Y} = \frac{1}{j\omega L}$
Capacitance C	$\boldsymbol{Z} = \frac{1}{j\omega C}$	$\boldsymbol{Y} = j\omega C$

Reactance X can be positive or negative. For inductive load, X is positive, and for capacitive load, X is negative.

Example 3.2

Determine the total impedance of a circuit in polar form at 60 [Hz] frequency, which contains 1 [H] inductor in series with a 100 [Ω] resistor.

Solution

The total circuit impedance is:

$$\mathbf{Z} = R + jX = R + j\omega L = R + j2\pi f L$$

Since $f = 60\ [\text{Hz}]$, we have:

$$\mathbf{Z} = R + j2\pi f L = 100 + j2\pi * 60 * 1$$
$$= 100 + j377 = 390.04 \angle 75.14^\circ\ [\Omega]$$

Example 3.3

Determine the total impedance of a circuit in polar form at 60 [Hz] frequency, which contains a 0.1 [H] inductor in series with a 1 [μF] capacitor and a 10 [Ω] resistor.

Solution

The total circuit impedance is:

$$\mathbf{Z} = R + jX = R + j\left(\omega L - \frac{1}{\omega C}\right) = R + j\left(2\pi f L - \frac{1}{2\pi f C}\right)$$

Since $f = 60\ [\text{Hz}]$, we have:

$$\mathbf{Z} = R + j\left(2\pi f L - \frac{1}{2\pi f C}\right) = 10 + j\left(2\pi * 60 * 0.1 - \frac{1}{2\pi * 60 * 10^{-6}}\right)$$
$$= 10 - j2614.88 = 2614.9 \angle -89.78^\circ\ [\Omega]$$

Example 3.4

A circuit contains a capacitor in series with a coil, which has an inductance of 2 [H] and a resistance of 10 [Ω]. The overall impedance of the circuit is purely resistive. Determine the value of the capacitance.

Solution

Since the overall circuit impedance is purely resistive, the capacitive reactance X_C cancels the inductive reactance X_L.

The reactance of the inductor is:

$$X_L = 2\pi * 60 * 2 = 754\ [\Omega].$$

The reactance of the capacitor therefore is –754 [Ω].

We then have:

$$X_C = -\frac{1}{\omega C}$$

$$C = -\frac{1}{\omega X_C} = -\frac{1}{2\pi f X_C} = -\frac{1}{2\pi * 60 * (-754)} = 3.51 \times 10^{-6}\,[\text{F}] = 3.51\ [\mu\text{F}]$$

Example 3.5

A current of 20 [A] flows through a load having an impedance of $(3 + j4)$ [Ω]. Determine the real and reactive powers supplied to the load.

Solution

The load impedance is:

$$\boldsymbol{Z} = R + jX = 3\ +\ j4\ [\Omega]$$

The real power supplied to the load is:

$$P = |\boldsymbol{I}|^2\,R = 20^2 \times 3 = 1200\ [\text{W}]$$

The reactive power supplied to the load is:

$$Q = |\boldsymbol{I}|^2\,X = 20^2 \times 4 = 1600\ [\text{VAR}]$$

Example 3.6

An inductive circuit load has a voltage of $120\angle 0°\ [\text{V}]$ across it, and the current flowing through it is $15\angle -60°\ [\text{A}]$. Determine the inductive circuit load resistance and inductance.

Solution

The inductive circuit load impedance is:

$$\boldsymbol{Z} = \frac{\boldsymbol{V}}{\boldsymbol{I}} = \frac{120\angle 0°}{15\angle -60°} = 8\angle 60° = 4 + j6.93\ [\Omega]$$

The resistance of the inductive circuit load is 4 [Ω].

The inductance of the load is:

$$L = \frac{6.93}{2\pi f} = \frac{6.93}{2\pi * 60} = 0.0184\ [\text{H}] = 18.40\ [\text{mH}]$$

3.6 POWER TRIANGLE

Real, reactive, and apparent powers supplied to a load are related by the power triangle as shown in Figure 3.5. Angle θ in the power triangle is called the impedance angle. Hypotenuse, adjacent side, and opposite side of the power triangle represent, respectively, apparent power S, real power P, and apparent power Q supplied to the load [10,40,43].

We thus have:

$$\cos\theta = \frac{P}{S} \quad \sin\theta = \frac{Q}{S} \quad \tan\theta = \frac{Q}{P} \tag{3.30}$$

The quantity $\cos\theta$ is called the power factor of the load. Power factor is defined as the fraction of apparent power S that is actually supplying power to the load [25,28].

$$\text{Power factor} = \text{PF} = \cos\theta \tag{3.31}$$

θ in the above expression is the impedance angle of the load as described previously.

We also have $\cos\theta = \cos(-\theta)$. The power factor produced by the impedance angle θ is therefore the same as the power factor produced by the impedance angle $-\theta$. Since it is not possible to determine directly from the power factor whether the load is inductive or capacitive, it is required to state whether the current is lagging or leading the voltage whenever the power factor is quoted.

The power triangle clarifies the relationship between P, Q, S, and PF, and it is also helpful in computing different power-related quantities if some of them are unknown. If any two of the quantities among P, Q, S, and PF are specified, the remaining two can be easily obtained from the power triangle.

Similar to the power triangle, the impedance triangle is constructed to describe the relationship between Z, R, and X [1]. The impedance triangle of a load is shown in Figure 3.6.

The reactive power Q supplied to a load is positive if $\boldsymbol{V}$ leads $\boldsymbol{I}$ and θ is positive, thus making $\sin\theta$ and $|\boldsymbol{V}||\boldsymbol{I}|\sin\theta$ positive. Similarly, the reactive power Q supplied to

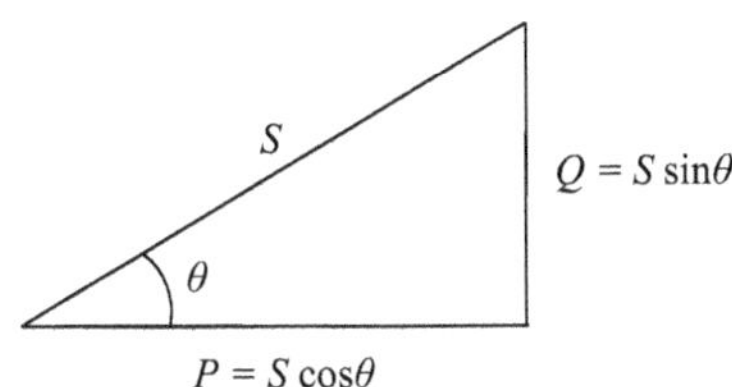

FIGURE 3.5 Power triangle.

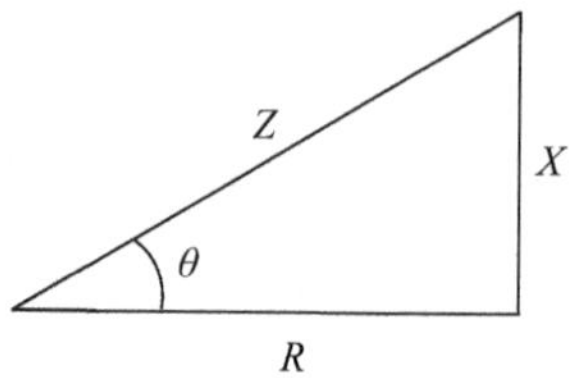

FIGURE 3.6 Impedance triangle of a load.

TABLE 3.2
Voltage-Current Phase Relations and Signs of Powers Consumed for Various Types of Loads

Load Type	Phase Relation between Voltage and Current	Power Absorbed by Load	
		P	Q
Purely resistive load	V and I in phase	$P>0$	$Q=0$
Purely inductive load	V leads I by 90°	$P=0$	$Q>0$
Purely capacitive load	V lags I by 90°	$P=0$	$Q<0$
Resistance and inductor in series or parallel	V leads I	$P>0$	$Q>0$
Resistance and capacitor in series or parallel	V lags I	$P>0$	$Q<0$

a load is negative if V lags I and θ is negative, thus making $\sin\theta$ and $|V||I|\sin\theta$ negative. The power factor is leading for inductive loads and lagging for capacitive loads. Table 3.2 shows the voltage-current phase relations and signs of powers consumed for various types of loads.

Example 3.7

Find the real and reactive powers supplied to a load having an impedance of $10\angle 30°\ [\Omega]$ when a voltage of $120\angle 0°\ [\text{V}]$ is applied across it. Also, determine the power factor of the load.

Solution

The current flowing through the load is:

$$\boldsymbol{I} = \frac{\boldsymbol{V}}{\boldsymbol{Z}} = \frac{120\angle 0°}{10\angle 30°} = 12\angle -30°\ [\text{A}]$$

The real power supplied to the load is:

$$P = |\boldsymbol{I}|^2 |\boldsymbol{Z}| \cos\theta = 12^2 \times 10 \times \cos 30° = 12^2 \times 10 \times 0.866 = 1247.04\ [\text{W}]$$

The reactive power supplied to the load is:

$$Q = |\boldsymbol{I}|^2 |\mathbf{Z}| \sin\theta = 12^2 \times 10 \times \sin 30° = 12^2 \times 10 \times 0.50 = 720\ [\text{VAR}]$$

The power factor of the load is:

$$\text{PF} = \cos\theta = \cos 30° = 0.866 \text{ lagging}$$

Since $\boldsymbol{V}$ leads $\boldsymbol{I}$, load is inductive, and therefore, the power factor is lagging.

Example 3.8

What is the reactive power supplied to an inductive load which has an apparent power input of 100 [kVA] and the real power absorbed by it is 60 [kW]? Also, determine the power factor of the load.

Solutizon

The reactive power supplied to the load is:

$$Q = \sqrt{S^2 - P^2} = \sqrt{100^2 - 60^2} = \sqrt{6400} = 80\ [\text{kVAR}]$$

The power factor of the load is:

$$\text{PF} = \cos\theta = \frac{P}{S} = \frac{60}{100} = 0.6 \text{ lagging}$$

Since the load is inductive, the power factor is lagging.

Example 3.9

Figure 3.7 shows an AC voltage source of $120\angle 0°$ [V] supplying power to a load having impedance $\mathbf{Z} = |\mathbf{Z}|\angle\theta = 10\angle -30°$ [Ω]. Determine the current $\boldsymbol{I}$ supplied to the load, load PF, and real power, reactive power, apparent power, and complex power supplied to the load.

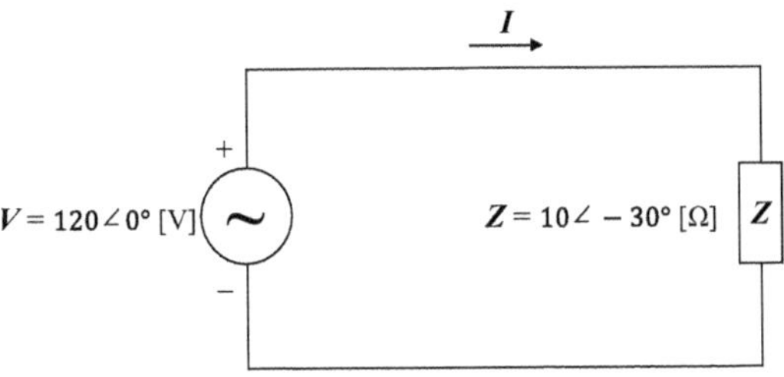

FIGURE 3.7 Example of an AC circuit.

Solution

The current supplied to the load is:

$$\boldsymbol{I} = \frac{\boldsymbol{V}}{\boldsymbol{Z}} = \frac{120\angle 0^\circ}{10\angle -30^\circ} = 12\angle 30^\circ \text{ [A]}$$

The load PF is:

$$\text{PF} = \cos\theta = \cos(-30^\circ) = 0.866 \text{ leading}$$

In a capacitive load, θ is negative, $\boldsymbol{I}$ leads $\boldsymbol{V}$, and PF is leading.

The real power supplied to the load is:

$$P = |\boldsymbol{V}||\boldsymbol{I}|\cos\theta = 120 \times 12 \times \cos(-30^\circ) = 1247 \text{ [W]}$$

The reactive power supplied to the load is:

$$Q = |\boldsymbol{V}||\boldsymbol{I}|\sin\theta = 120 \times 12 \times \sin(-30^\circ) = -720 \text{ [VAR]}$$

The apparent power supplied to the load is:

$$S = |\boldsymbol{V}||\boldsymbol{I}| = 120 \times 12 = 1440\text{[VA]}$$

The complex power supplied to the load is:

$$\begin{aligned} \boldsymbol{S} = \boldsymbol{V}\boldsymbol{I}^* &= 120\angle 0^\circ \times (12\angle 30^\circ)^* \\ &= 1440\angle -30^\circ\text{[VA]} \\ &= 1247 - j720\text{[VA]} \\ &= P + jQ \end{aligned}$$

3.7 CONSERVATION OF POWER IN AC CIRCUITS

AC circuits are subject to the principle of conservation of power. We consider the parallel circuit shown in Figure 3.8 to verify the conservation of power in AC circuits. In this circuit, an AC voltage source V supplies power to three loads connected in parallel.

In the circuit below, an AC voltage source $\boldsymbol{V}$ supplies the total power $\boldsymbol{S}$ and the total current $\boldsymbol{I}$. The currents supplied to the three loads 1, 2, and 3 connected in parallel are $\boldsymbol{I}_1$, $\boldsymbol{I}_2$, and $\boldsymbol{I}_3$, respectively.

We therefore have:

$$\boldsymbol{I} = \boldsymbol{I}_1 + \boldsymbol{I}_2 + \boldsymbol{I}_3 \tag{3.32}$$

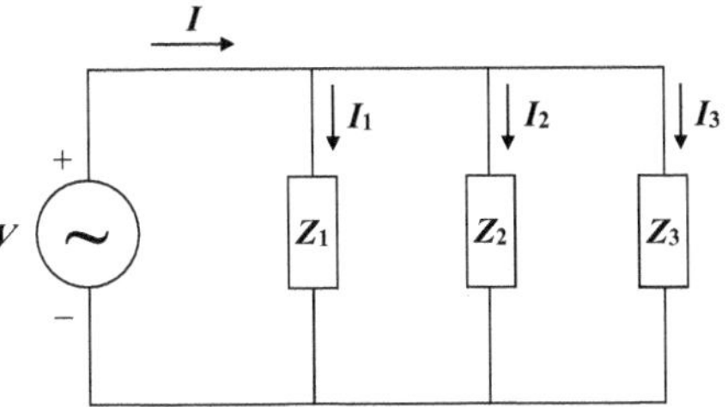

FIGURE 3.8 Parallel AC circuit.

We further have:

$$\begin{aligned} \boldsymbol{S} &= \boldsymbol{V}\boldsymbol{I}^* = \boldsymbol{V}\left[\boldsymbol{I}_1 + \boldsymbol{I}_2 + \boldsymbol{I}_3\right]^* \\ &= \boldsymbol{V}\boldsymbol{I}_1 + \boldsymbol{V}\boldsymbol{I}_2 + \boldsymbol{V}\boldsymbol{I}_3 \\ &= \boldsymbol{S}_1 + \boldsymbol{S}_2 + \boldsymbol{S}_3 \end{aligned} \tag{3.33}$$

$\boldsymbol{S}_1$, $\boldsymbol{S}_2$, and $\boldsymbol{S}_3$ in the above equation are complex powers supplied to the three loads 1, 2, and 3, which are connected in parallel.

We find that the total power supplied by the source equals the total power received by the loads when loads are connected in parallel in an AC circuit. The principle of conservation of power is therefore verified for parallel AC circuits.

We next consider the series circuit shown in Figure 3.9. In this circuit, an AC voltage source $\boldsymbol{V}$ supplies the total power $\boldsymbol{S}$ and the total current $\boldsymbol{V}$ to three loads connected in series. The voltages across the three series-connected loads 1, 2, and 3 are $\boldsymbol{V}_1$, $\boldsymbol{V}_2$, and $\boldsymbol{V}_3$, respectively.

We therefore have:

$$\begin{aligned} \boldsymbol{S} &= \boldsymbol{V}\boldsymbol{I}^* = \left[\boldsymbol{V}_1 + \boldsymbol{V}_2 + \boldsymbol{V}_3\right]\boldsymbol{I}^* \\ &= \boldsymbol{V}_1\boldsymbol{I}^* + \boldsymbol{V}_2\boldsymbol{I}^* + \boldsymbol{V}_3\boldsymbol{I}^* \\ &= \boldsymbol{S}_1 + \boldsymbol{S}_2 + \boldsymbol{S}_3 \end{aligned} \tag{3.34}$$

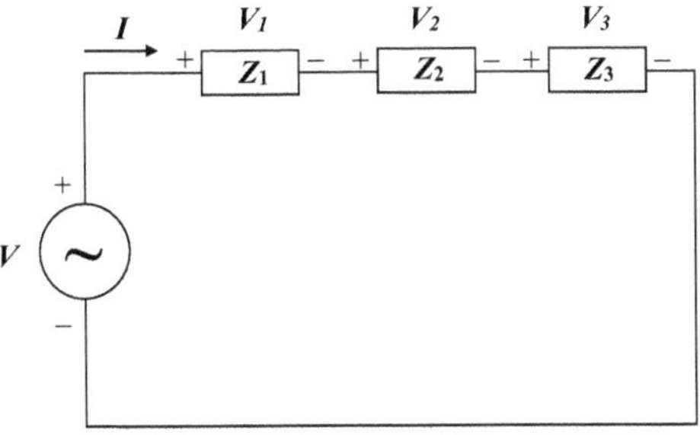

FIGURE 3.9 Series AC circuit.

S_1, S_2, and S_3 in the above equation are complex powers supplied to the three loads 1, 2, and 3, which are connected in series.

We find that the total power supplied by the source equals the total power received by the loads when loads are connected in series in an AC circuit. The principle of conservation of power is therefore verified for series AC circuits.

From the above description, we can draw the general conclusion that power is conserved in AC circuits irrespective of whether the loads are connected in series, in parallel, or in any general combination, and the total power which the source supplies equals the total power received by the loads.

For the case of an AC voltage source connected to N loads in an AC circuit, we therefore have the following result:

$$S = S_1 + S_2 + \cdots + S_N \tag{3.35}$$

Furthermore, we can easily verify from the above result that complex, real, and reactive powers supplied by the source equal the respective sums of complex, real, and reactive powers received by the individual loads.

Example 3.10

Figure 3.10 shows an AC voltage source of $600\angle 0°$ [V] supplying power to two loads $10 + j0$ [Ω] and $5 + j3$ [Ω] connected in parallel. Determine the complex powers absorbed by the two load branches. Also, verify the law of conservation of AC power for the circuit.

Solution

The complex power supplied to load 1 is:

$$S_1 = \frac{|V^2|}{Z_1^*} = \frac{(600)^2}{(10 + j0)^*} = (36000 + j0)\ [\text{VA}]$$

The complex power supplied to load 2 is:

$$S_2 = \frac{|V^2|}{Z_2^*} = \frac{(600)^2}{(5 + j3)^*} = (52941 + j31765)\ [\text{VA}]$$

The sum of complex powers supplied to loads 1 and 2 is:

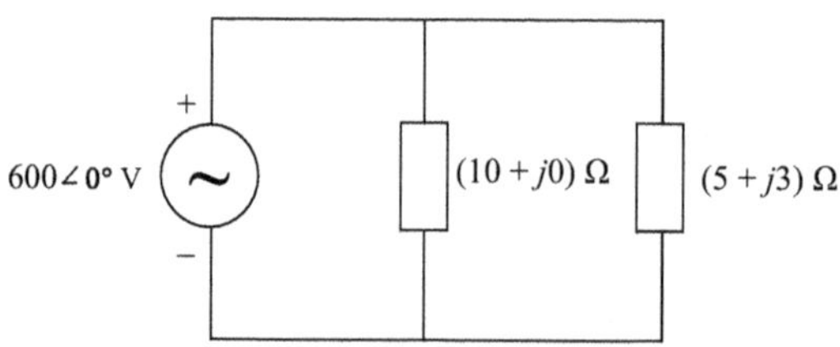

FIGURE 3.10 Example of an AC circuit.

$$S_1 + S_2 = (36000 + j0) + (52941 + j31765) = (88941 + j31765)\ [\text{VA}]$$

The equivalent impedance of the parallel combination of loads 1 and 2 is:

$$Z = \frac{Z_1 \times Z_2}{Z_1 + Z_2} = \frac{(10 + j0) \times (5 + j3)}{(10 + j0) + (5 + j3)} = 3.81\angle 19.65°\ [\Omega]$$

The complex power supplied to equivalent load impedance is:

$$S = \frac{|V^2|}{Z^*} = \frac{(600)^2}{(3.81\angle 19.65°)^*} = (88941 + j31765)\ [\text{VA}]$$

We thus have:

$$S = S_1 + S_2$$

3.8 SUMMARY

Single-phase AC circuits have been described in this chapter. AC is an electric current which periodically reverses direction and has its magnitude varying continuously with time. The usual waveform of AC is a sine wave. Circuits driven by sinusoidal current or voltage sources are called AC circuits. Phasor analysis is an efficient technique most commonly employed for analyzing AC circuits.

The three kinds of power in an AC circuit are real power, reactive power, and apparent power. These powers are related by the power triangle. The principle of conservation of power applies to AC circuits. When loads are connected in series, in parallel, or in any general combination in an AC circuit, the total power supplied by the source equals the total power received by the loads.

REVIEW QUESTIONS

3.1 Why is alternating current preferred for the supply of power in electrical power systems?
3.2 Distinguish between real power, reactive power, and apparent power in an AC circuit.
3.3 Derive expressions for real power, reactive power, and apparent power consumed in an AC circuit.
3.4 What does the term power factor of a load mean?
3.5 Describe why the power factor is leading for inductive loads and lagging for capacitive loads.
3.6 Prove that real power consumed in a pure inductive load is zero.
3.7 Prove that real power consumed in a pure capacitive load is zero.
3.8 What are the reasons for the use of phasor analysis in AC circuit calculations?
3.9 Describe how real power, reactive power, and apparent power supplied to a load are related by the power triangle.
3.10 Describe what is an impedance triangle.
3.11 Prove that power is conserved in a series AC circuit.
3.12 Prove that power is conserved in a parallel AC circuit.

4 AC Circuit Fundamentals
Three-Phase Circuits

4.1 INTRODUCTION

The electric power generated in power plants by three-phase synchronous generators is of three-phase alternating current (AC) form mainly due to the relatively simple process of transformation of the generated voltage to a higher transmission voltage level for economic and efficient power transmission with lower losses, constancy of power supplied in three-phase systems, and availability of low-cost three-phase and single-phase AC motors to provide the motive power.

A three-phase voltage source has three sinusoidal voltage sources of equal magnitude which differ in phase angle by 120° from each other. Assuming *abc* phase sequence, the voltage source in phase *b* lags the voltage source in phase *a* by a phase angle of 120°, whereas the voltage source in phase *c* lags the voltage source in phase *b* by a phase angle of 120°. During proper operation of three-phase power systems, voltages and currents in the three phases are balanced with a progressive phase difference of 120°.

Analysis of a three-phase system is simplified by the use of its per-phase equivalent circuit for the determination of system voltages and currents in one of the three phases. From the system voltage and current results obtained for one phase, it is possible to easily determine voltages and currents in the other two phases.

This chapter describes relationships between voltages and currents in balanced three-phase systems, which can have voltage sources and loads connected in wye or delta. The power relationships in a three-phase system and analysis of three-phase systems using the per-phase equivalent circuit method are also described.

4.2 VOLTAGES AND CURRENTS IN THREE-PHASE SYSTEMS

A three-phase (3-ϕ) synchronous generator is designed to generate balanced three-phase voltages.

The generated voltages are of equal magnitude; however, they differ in phase from each other by 120°. Each of the three-phase terminals of the three-phase synchronous generator can be connected to a single-phase load $\mathbf{Z} = |\mathbf{Z}| \angle \theta$ for supplying power as shown in Figure 4.1. The single-phase loads in each of the three phases must be equal for the balanced operation of a three-phase system. Figure 4.2 shows the waveforms of the balanced three-phase voltages.

The voltages generated in the three phases of a three-phase system are represented using the following equations:

DOI: 10.1201/9781003432340-4

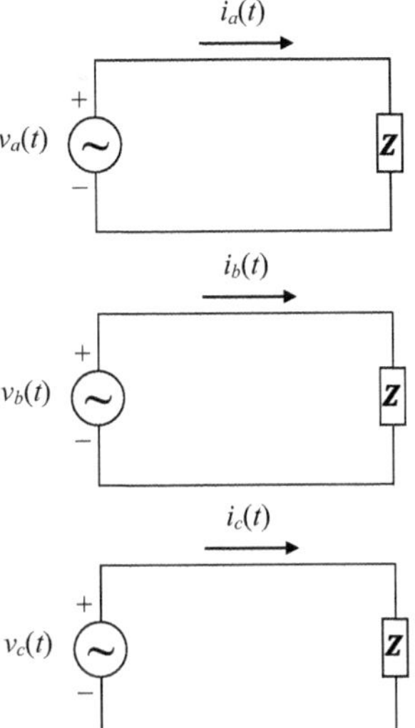

FIGURE 4.1 Three-phase system voltage sources connected separately to balanced single-phase loads.

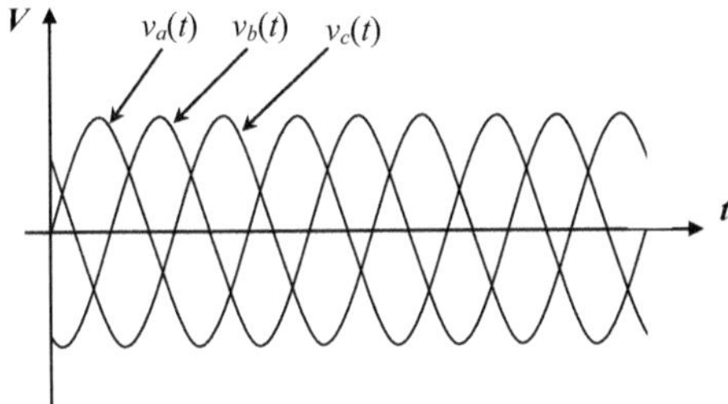

FIGURE 4.2 Waveforms of balanced three-phase voltages.

$$\begin{aligned} v_a &= \sqrt{2}|\boldsymbol{V}|\sin\omega t \ [\mathrm{V}] \\ v_b &= \sqrt{2}|\boldsymbol{V}|\sin(\omega t - 120°) \ [\mathrm{V}] \\ v_c &= \sqrt{2}|\boldsymbol{V}|\sin(\omega t - 240°) \ [\mathrm{V}] \end{aligned} \tag{4.1}$$

We can rewrite the above equations in phasor form (Figure 4.3) as:

$$\begin{aligned} \boldsymbol{V_a} &= |\boldsymbol{V}| \angle 0° \ [\mathrm{V}] \\ \boldsymbol{V_b} &= |\boldsymbol{V}| \angle -120° \ [\mathrm{V}] \\ \boldsymbol{V_c} &= |\boldsymbol{V}| \angle -240° \ [\mathrm{V}] \end{aligned} \tag{4.2}$$

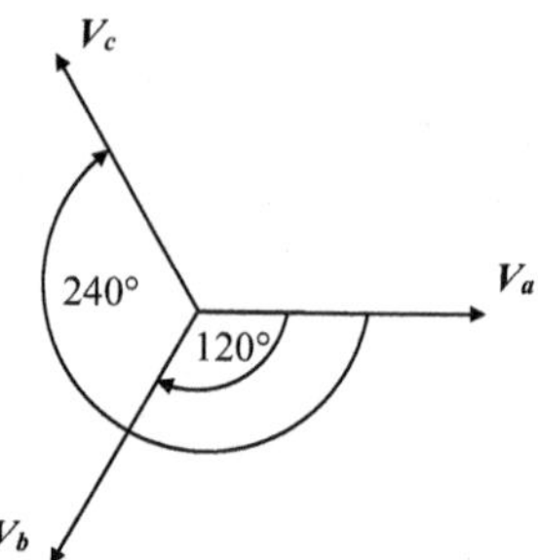

FIGURE 4.3 Phasor voltages in a three-phase system.

The generated three-phase voltages shown in Figures 4.2 and 4.3 above have the phase sequence *abc*, since voltages in the three phases reach their maximum value in that order. Another phase sequence *acb* is also possible, in which voltages in the three phases reach their maximum values in the order *a*, *c*, *b*. Results obtained for *abc* phase sequence are applicable to *acb* phase sequence after slight phase order correction. In this reference, only phase sequence *abc* is used.

The currents supplied to the three single-phase loads in a three-phase system are calculated as follows:

$$\begin{aligned}
\boldsymbol{I}_a &= \frac{\boldsymbol{V}_a}{\boldsymbol{Z}} = \frac{|\boldsymbol{V}|\angle 0^\circ}{|\boldsymbol{Z}|\angle\theta} = |\boldsymbol{I}|\angle-\theta \\
\boldsymbol{I}_b &= \frac{\boldsymbol{V}_b}{\boldsymbol{Z}} = \frac{|\boldsymbol{V}|\angle-120^\circ}{|\boldsymbol{Z}|\angle\theta} = |\boldsymbol{I}|\angle-(120^\circ+\theta) \\
\boldsymbol{I}_c &= \frac{\boldsymbol{V}_c}{\boldsymbol{Z}} = \frac{|\boldsymbol{V}|\angle-240^\circ}{|\boldsymbol{Z}|\angle\theta} = |\boldsymbol{I}|\angle-(240^\circ+\theta)
\end{aligned} \tag{4.3}$$

where

$$|\boldsymbol{I}| = \frac{|\boldsymbol{V}|}{|\boldsymbol{Z}|} \tag{4.4}$$

The three voltage sources and loads in the three-phase system above can have their negative ends connected so that the three load currents have a common return conductor as shown in Figure 4.4. The resulting current in the neutral return wire is the sum of load currents in the three phases calculated above:

$$\begin{aligned}
\boldsymbol{I}_n &= \boldsymbol{I}_a + \boldsymbol{I}_b + \boldsymbol{I}_c = |\boldsymbol{I}|\angle-\theta + |\boldsymbol{I}|\angle-(120^\circ+\theta) + |\boldsymbol{I}|\angle-(240^\circ+\theta) \\
&= |\boldsymbol{I}|\left[\cos\theta + \cos(120^\circ+\theta) + \cos(240^\circ+\theta)\right] \\
&\quad - j|\boldsymbol{I}|\left[\sin\theta + \sin(120^\circ+\theta) + \sin(240^\circ+\theta)\right] \\
&= 0 - j0 \\
&= 0
\end{aligned} \tag{4.5}$$

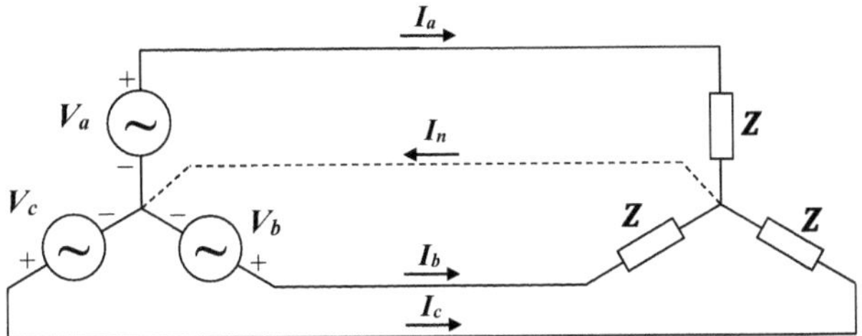

FIGURE 4.4 The three circuits connected together with a common neutral return conductor.

The current in the neutral wire of the above connection of three single-phase sources and loads is zero. The neutral wire is therefore not necessary and can be eliminated without affecting the operation of the above circuit [28]. We therefore require only three wires in the above interconnection of three single-phase systems if the neutral wire is eliminated and four wires if the neutral wire is maintained to supply the same amount of power as that in three single-phase systems working separately, which require six wires.

The three-phase generator can supply power to both single-phase loads connected separately in each of the three phases and three-phase loads. Loads in the three phases are kept balanced in the case of both single-phase loads and three-phase loads for the balanced operation of a three-phase power system [6,12].

Example 4.1

A balanced three-phase voltage source with an *abc* phase sequence has phase voltage $\boldsymbol{V_a} = 120\angle 30°\ [\mathrm{A}]$. Determine $\boldsymbol{V_b}$ and $\boldsymbol{V_c}$.

Solution

Since the circuit is balanced, all three-phase voltages have the same magnitude of $120\ [\mathrm{V}]$. Furthermore, since the phase sequence is *abc*, $\boldsymbol{V_b}$ lags $\boldsymbol{V_a}$ by 120°. Similarly, $\boldsymbol{V_c}$ lags $\boldsymbol{V_b}$ by 120°.

We therefore have:

$$\boldsymbol{V_b} = 120\angle(30° - 120°) = 120\angle -90°\ [\mathrm{V}]$$

$$\boldsymbol{V_c} = 120\angle(-90° - 120°) = 120\angle -210°\ [\mathrm{V}]$$

Example 4.2

A balanced three-phase circuit with an *abc* phase sequence has line current $\boldsymbol{I_b} = 50\angle 60°\ [\mathrm{A}]$. Determine the other line currents.

Solution

Since the circuit is balanced, all three line currents have the same magnitude of $50\ [\mathrm{A}]$. Furthermore, since the phase sequence is *abc*, $\boldsymbol{I_a}$ leads $\boldsymbol{I_b}$ by 120°. Also, $\boldsymbol{I_c}$ lags $\boldsymbol{I_b}$ by 120°.

We therefore have:

$$I_a = 50\angle(60° + 120°) = 50\angle 180° \text{ [A]}$$

$$I_c = 50\angle(60° - 120°) = 50\angle -60° \text{ [A]}$$

4.3 WYE AND DELTA CONNECTIONS OF THREE-PHASE VOLTAGE SOURCES AND THREE-PHASE LOADS

The three-phase voltage sources can be either wye-connected or delta-connected as shown in Figure 4.5. The three-phase loads can be similarly wye-connected or delta-connected as shown in Figure 4.6. Any number of wye- or delta-connected generators and loads can be present in a power system.

Three-phase voltage sources are rarely connected in delta configuration, since circulating current flows in the delta if the three-phase voltages are slightly unbalanced. Similarly, balanced delta-connected loads are more common than balanced wye-connected loads, since connection or disconnection of delta-connected loads between phases is easier, and connection or disconnection of wye-connected loads is more difficult as the neutral is not accessible in some cases.

Since three-phase voltage sources are mainly connected in wye, and loads are both wye- and delta-connected, we shall study only the following two connections of three-phase voltage sources and loads:

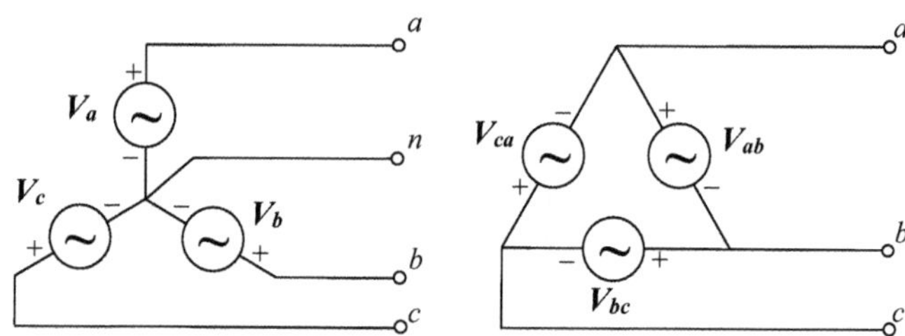

FIGURE 4.5 Wye and delta connections of three-phase voltage sources.

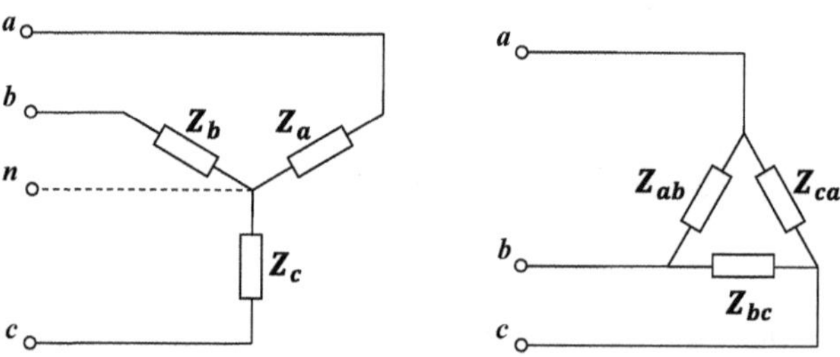

FIGURE 4.6 Wye and delta connections of three-phase loads.

1. A wye-connected source supplying power to a wye-connected load.
2. A wye-connected source supplying power to a delta-connected load.

4.4 BALANCED WYE-WYE CONNECTION

The balanced wye-wye circuit connection in which a balanced wye-connected source supplies power to a balanced wye-connected load is shown in Figure 4.7. The impedance of the line connecting the source to the load is neglected in the analysis since it is much less than the load impedance.

We have:

$$\begin{aligned} \boldsymbol{V}_a &= |\boldsymbol{V}|\angle 0° \text{ [V]} \\ \boldsymbol{V}_b &= |\boldsymbol{V}|\angle -120° \text{ [V]} \\ \boldsymbol{V}_c &= |\boldsymbol{V}|\angle -240° \text{ [V]} \end{aligned} \tag{4.6}$$

We calculate the relationship between phase voltages and line voltages as follows:

$$\begin{aligned} \boldsymbol{V}_{ab} = \boldsymbol{V}_{AB} &= \boldsymbol{V}_a - \boldsymbol{V}_b \\ &= |\boldsymbol{V}|\angle 0° - |\boldsymbol{V}|\angle -120° \\ &= \sqrt{3}|\boldsymbol{V}|\angle 30° \text{ [V]} \end{aligned}$$

$$\begin{aligned} \boldsymbol{V}_{bc} = \boldsymbol{V}_{BC} &= \sqrt{3}|\boldsymbol{V}|\angle -90° \text{ [V]} \\ \boldsymbol{V}_{ca} = \boldsymbol{V}_{CA} &= \sqrt{3}|\boldsymbol{V}|\angle -210° \text{ [V]} \end{aligned} \tag{4.7}$$

The three line voltages therefore form a balanced three-phase set, and they are advanced in phase by 30° from the phase voltages.

In terms of magnitude, we can therefore write:

$$|V_{LL}| = \sqrt{3}|\boldsymbol{V}_f| = \sqrt{3}|\boldsymbol{V}| \tag{4.8}$$

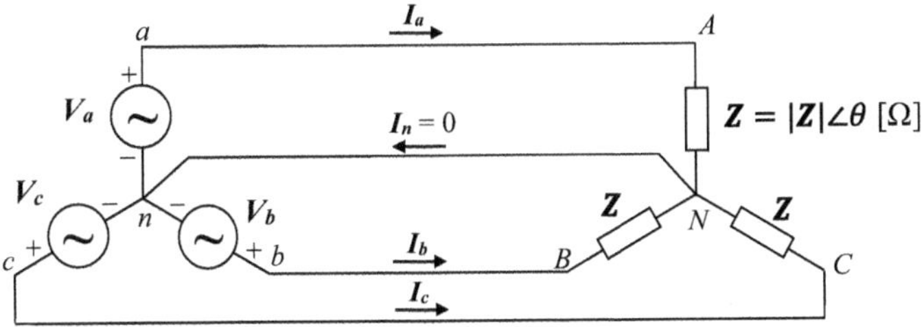

FIGURE 4.7 Balanced wye-wye connection.

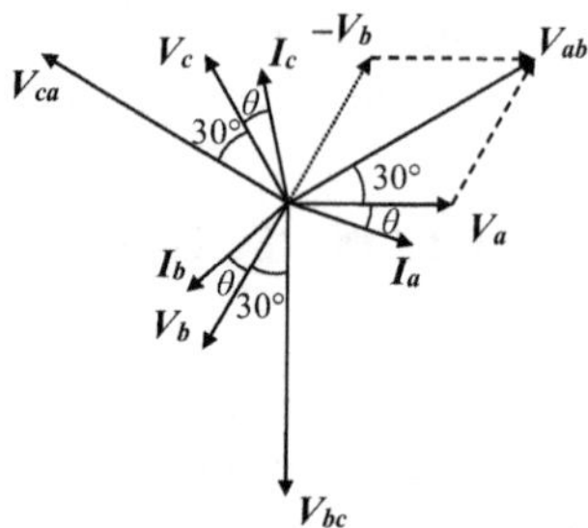

FIGURE 4.8 Relationships between various voltage and current phasors in wye-wye connection.

We also have:

$$\boldsymbol{V_{ab}} + \boldsymbol{V_{bc}} + \boldsymbol{V_{ca}} = \sqrt{3}|V|\angle 30° + \sqrt{3}|V|\angle -90° + \sqrt{3}|V|\angle -210° = 0 \quad (4.9)$$

The sum of the three line voltage phasors in the above balanced three-phase system is zero as expected.

We next compute the line currents as follows:

$$\boldsymbol{I_a} = \frac{\boldsymbol{V_a}}{\boldsymbol{Z}} = \frac{|\boldsymbol{V}|}{|\boldsymbol{Z}|}\angle -\theta$$

$$\boldsymbol{I_b} = \frac{\boldsymbol{V_b}}{\boldsymbol{Z}} = \frac{|\boldsymbol{V}|}{|\boldsymbol{Z}|}\angle -(120° + \theta) \quad (4.10)$$

$$\boldsymbol{I_c} = \frac{\boldsymbol{V_c}}{\boldsymbol{Z}} = \frac{|\boldsymbol{V}|}{|\boldsymbol{Z}|}\angle -(240° + \theta)$$

$$\boldsymbol{I_a} + \boldsymbol{I_b} + \boldsymbol{I_c} = \frac{|\boldsymbol{V}|}{|\boldsymbol{Z}|}\angle -\theta + \frac{|\boldsymbol{V}|}{|\boldsymbol{Z}|}\angle -(120° + \theta) + \frac{|\boldsymbol{V}|}{|\boldsymbol{Z}|}\angle -(240° + \theta) = 0 \quad (4.11)$$

The line currents therefore form a balanced three-phase set. There is no current in the neutral wire, and n and N have the same potential. It can also be easily observed that phase currents and line currents are identical in wye-wye connection. The relationships between various voltage and current phasors in wye-wye connection are shown in Figure 4.8.

Example 4.3

A balanced Y-connected source with $\boldsymbol{V_a} = 120\angle 0°$ [V] is supplying power to a Y-connected balanced per-phase load of $20\angle 30°$ [Ω]. Determine the line currents and current in the neutral wire.

Solution

The phase voltage $\boldsymbol{V}_a$ is:

$$\boldsymbol{V}_a = 120\angle 0^\circ \text{ [V]}$$

The Y-connected balanced per-phase load impedance is:

$$\boldsymbol{Z} = 20\angle 30^\circ \text{ [}\Omega\text{]}$$

The line currents therefore are:

$$\boldsymbol{I}_a = \frac{\boldsymbol{V}_a}{\boldsymbol{Z}} = \frac{120\angle 0^\circ}{20\angle 30^\circ} = 6\angle -30^\circ \text{ [A]}$$

$$\boldsymbol{I}_b = \boldsymbol{I}_a\angle -120^\circ = 6\angle -150^\circ \text{ [A]}$$

$$\boldsymbol{I}_c = \boldsymbol{I}_b\angle -120^\circ = 6\angle -270^\circ = 6\angle 90^\circ \text{ [A]}$$

Current in the neutral wire is:

$$\boldsymbol{I}_n = \boldsymbol{I}_a + \boldsymbol{I}_b + \boldsymbol{I}_c = 6\angle -30^\circ + 6\angle -150^\circ + 6\angle 90^\circ = 0 \text{ [A]}$$

4.5 BALANCED WYE-DELTA CONNECTION

The balanced wye-delta circuit connection in which a balanced wye-connected source supplies power to a balanced delta-connected load is shown in Figure 4.9. Similar to the balanced wye-delta connection, the impedance of the line connecting the source to the load is neglected in the analysis since it is much less than the load impedance.

We have:

$$\begin{aligned} \boldsymbol{V}_a &= |\boldsymbol{V}|\angle 0^\circ \text{ [V]} \\ \boldsymbol{V}_b &= |\boldsymbol{V}|\angle -120^\circ \text{ [V]} \\ \boldsymbol{V}_c &= |\boldsymbol{V}|\angle -240^\circ \text{ [V]} \end{aligned} \tag{4.6}$$

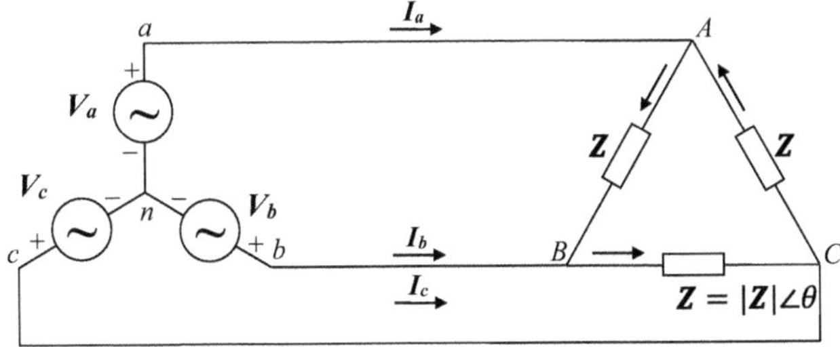

FIGURE 4.9 Balanced wye-delta connection.

We calculate the relationship between phase voltages and line voltages as follows:

$$\begin{aligned}
V_{ab} &= V_a - V_b = |V|\angle 0^\circ - |V|\angle -120^\circ = \sqrt{3}|V|\angle 30^\circ = V_{AB} \\
V_{bc} &= \sqrt{3}|V|\angle -90^\circ = V_{BC} \\
V_{ca} &= \sqrt{3}|V|\angle -210^\circ = V_{CA}
\end{aligned} \tag{4.12}$$

We next compute the load phase currents as follows:

$$\begin{aligned}
I_{AB} &= \frac{V_{AB}}{Z} \\
I_{BC} &= \frac{V_{BC}}{Z} = I_{AB}\angle -120^\circ \\
I_{CA} &= \frac{V_{CA}}{Z} = I_{AB}\angle -240^\circ
\end{aligned} \tag{4.13}$$

The line currents in wye-delta connection are calculated as follows:

$$\begin{aligned}
I_a &= I_{AB} - I_{CA} = I_{AB}\left[1 - 1\angle -240^\circ\right] = \sqrt{3}I_{AB}\angle -30^\circ \\
I_b &= \sqrt{3}I_{AB}\angle -150^\circ \\
I_c &= \sqrt{3}I_{AB}\angle -270^\circ
\end{aligned} \tag{4.14}$$

$$I_a + I_b + I_c = 0 \tag{4.15}$$

In terms of magnitudes, we can write:

$$|I_\phi| = |I| = |I_{AB}| = |I_{BC}| = |I_{CA}| \tag{4.16}$$

$$|I_L| = |I_a| = |I_b| = |I_c| \tag{4.17}$$

$$|I_L| = \sqrt{3}|I_\phi| \tag{4.18}$$

The line currents and phase currents form a balanced three-phase set. The line current magnitudes are $\sqrt{3}$ times the phase current magnitudes, and line currents lag corresponding phase currents by 30°.

The relationships between various voltage and current phasors in wye-delta connection are shown in Figure 4.10.

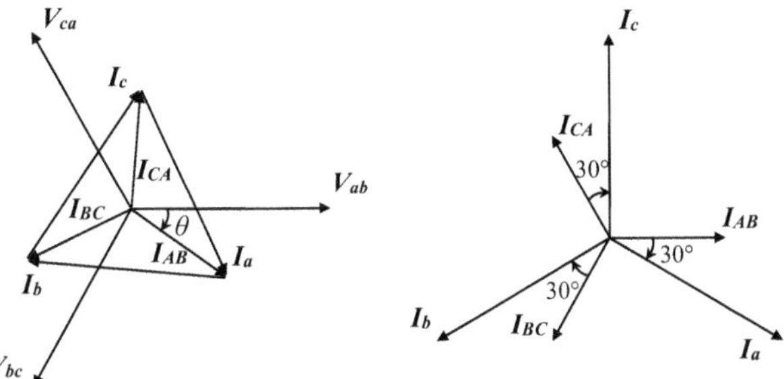

FIGURE 4.10 Relationships between various voltage and current phasors in wye-delta connection.

Example 4.4

A balanced Y-connected source with $V_a = 120\angle 0°$ [V] is supplying power to a Δ-connected balanced per-phase load of $10\angle 60°$ [Ω]. Determine the phase currents and line currents.

Solution

The phase voltage V_a is:

$$V_a = 120\angle 0° \text{ [V]}$$

The line voltage therefore is:

$$V_{ab} = V_{AB} = \sqrt{3}V_a\angle 30° = \sqrt{3}\times 120\ \angle(0°+30°) = 207.84\angle 30° \text{ [V]}$$

The balanced delta-connected load impedance is:

$$Z = 10\angle 60° \text{ [Ω]}$$

The phase currents therefore are:

$$I_{AB} = \frac{V_{AB}}{Z} = \frac{207.84\angle 30°}{10\angle 60°} = 20.78\angle -30° \text{ [A]}$$

$$I_{BC} = I_{AB}\angle -120° = 20.78\angle -150° \text{ [A]}$$

$$I_{CA} = I_{BC}\angle -120° = 20.78\angle -270° = 20.78\angle 90° \text{ [A]}$$

The line currents calculated from the above phase currents are:

$$I_a = \sqrt{3}I_{AB}\ \angle\ -30° = \sqrt{3}\times 20.78\angle -(30°+30°) = 36\angle -60° \text{ [A]}$$

$$I_b = I_a\angle -120° = 36\angle -180° \text{ [A]}$$

$$I_c = I_b\angle -120° = 36\angle -300° = 36\angle 60° \text{ [A]}$$

4.6 DELTA-WYE LOAD CONVERSION

A delta-connected load can be converted to an equivalent wye-connected load using the delta-wye conversion. Figure 4.11 shows balanced delta-connected impedances each of value $\mathbf{Z}_\Delta$ along with equivalent balanced wye-connected impedances each of value $\mathbf{Z}_Y$.

We have the following delta-wye transformation formula for converting delta-connected load impedance to equivalent wye-connected load impedance:

$$\mathbf{Z}_Y = \frac{\mathbf{Z}_\Delta}{3} \quad (4.19)$$

After this transformation, a wye-delta system with a balanced wye-connected source supplying power to a balanced delta-connected load can be converted to an equivalent wye-wye system with a balanced wye-connected source supplying power to a balanced wye-connected load. It is then possible to perform single-phase circuit analysis on a wye-wye system as described later.

Example 4.5

A balanced Δ-connected load has a per-phase impedance of $30\angle 60° \text{ [}\Omega\text{]}$. Determine the equivalent balanced *Y*-connected per-phase load impedance.

Solution

The Δ-connected balanced per-phase load impedance is:

$$\mathbf{Z}_\Delta = 30\angle 60° \text{ [}\Omega\text{]}$$

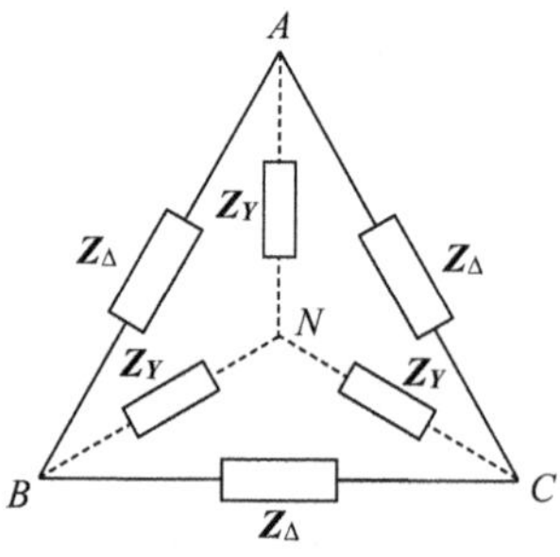

FIGURE 4.11 Delta-connected impedances and their equivalent wye-connected impedances.

The equivalent Y-connected balanced per-phase load impedance therefore is:

$$\boldsymbol{Z}_Y = \frac{\boldsymbol{Z}_\Delta}{3} = \frac{30\angle 60°}{3} = 10\angle 60° \ [\Omega]$$

4.7 POWER RELATIONSHIPS IN A THREE-PHASE CIRCUIT

We consider a balanced wye-connected load having phase impedance $\boldsymbol{Z} = |\boldsymbol{Z}|\angle\theta$ receiving power from a balanced three-phase voltage source with the following phase voltages:

$$\begin{aligned} v_a &= \sqrt{2}|\boldsymbol{V}|\sin\omega t \\ v_b &= \sqrt{2}|\boldsymbol{V}|\sin(\omega t - 120°) \\ v_c &= \sqrt{2}|\boldsymbol{V}|\sin(\omega t - 240°) \end{aligned} \tag{4.6}$$

The phase currents flowing in the loads are:

$$\begin{aligned} i_a &= \sqrt{2}\frac{|\boldsymbol{V}|}{|\boldsymbol{Z}|}\sin(\omega t - \theta) = \sqrt{2}|\boldsymbol{I}|\sin(\omega t - \theta) \\ i_b &= \sqrt{2}\frac{|\boldsymbol{V}|}{|\boldsymbol{Z}|}\sin(\omega t - 120° - \theta) = \sqrt{2}|\boldsymbol{I}|\sin(\omega t - 120° - \theta) \\ i_c &= \sqrt{2}\frac{|\boldsymbol{V}|}{|\boldsymbol{Z}|}\sin(\omega t - 240° - \theta) = \sqrt{2}|\boldsymbol{I}|\sin(\omega t - 240° - \theta) \end{aligned} \tag{4.20}$$

where

$$|\boldsymbol{I}| = \frac{|\boldsymbol{V}|}{|\boldsymbol{Z}|} \tag{4.4}$$

We next compute instantaneous power supplied to the loads in each of the three phases:

$$\begin{aligned} p_a &= v_a i_a = 2|\boldsymbol{V}||\boldsymbol{I}|\sin\omega t\sin(\omega t - \theta) \\ p_b &= v_b i_b = 2|\boldsymbol{V}||\boldsymbol{I}|\sin(\omega t - 120°)\sin(\omega t - 120° - \theta) \\ p_c &= v_c i_c = 2|\boldsymbol{V}||\boldsymbol{I}|\sin(\omega t - 240°)\sin(\omega t - 240° - \theta) \end{aligned} \tag{4.21}$$

We simplify the above instantaneous power equations using knowledge of trigonometry and obtain after simplification:

$$p_a = |V||I|\left[\cos\theta - \cos(2\omega t - \theta)\right]$$

$$p_b = |V||I|\left[\cos\theta - \cos(2\omega t - 240^\circ - \theta)\right] \tag{4.22}$$

$$p_c = |V||I|\left[\cos\theta - \cos(2\omega t - 480^\circ - \theta)\right]$$

Instantaneous power supplied to loads in each phase has a constant component and a pulsating component. We next sum the powers supplied to the loads connected to each of the three phases to obtain the total power supplied to the entire three-phase load:

$$\begin{aligned} p_{3\phi} = p_a + p_b + p_c &= 3|V||I|\cos\theta \\ &-|V||I|\left[\cos(2\omega t - \theta) + \cos(2\omega t - 240^\circ - \theta) + \cos(2\omega t - 480^\circ - \theta)\right] \\ &= 3|V||I|\cos\theta - |V||I| \times 0 \\ &= 3|V||I|\cos\theta \end{aligned} \tag{4.23}$$

In the above summation, the pulsating components in each of the three load currents cancel each other as they are 120° out of phase with each other. Total power supplied to a balanced three-phase load is consequently constant at all times, and it does not pulsate with time as the powers in each of the three individual phases. Constant instantaneous power supplied by three-phase power systems is a main advantage in comparison with single-phase power systems [7,12,14,24]. This result is applicable to all balanced three-phase loads irrespective of whether the loads are wye-connected or delta-connected.

Total reactive power supplied to a balanced three-phase load is similarly obtained from the following relationship:

$$Q_{3\phi} = 3|V||I|\sin\theta \tag{4.24}$$

Finally, total apparent power supplied to a balanced three-phase load is obtained from the following relationship:

$$S_{3\phi} = 3|V||I| \tag{4.25}$$

We next determine real power, reactive power, and apparent power relationships in terms of line quantities.

In wye-connected loads, we have:

$$|I_L| = |I_\phi| = |I| \tag{4.26}$$

$$|V_{LL}| = \sqrt{3}|V_\phi| = \sqrt{3}|V| \tag{4.27}$$

$$|V_\phi| = |V| = \frac{|V_{LL}|}{\sqrt{3}}$$

$$P_{3\phi} = 3|V||I|\cos\theta = 3|V_\phi||I_\phi| = \cos\theta = 3\frac{|V_{LL}|}{\sqrt{3}}|I_L|\cos\theta = \sqrt{3}|V_{LL}||I_L|\cos\theta \quad (4.28)$$

In delta-connected loads, we have:

$$|V_{LL}| = |V_\phi| = |V| \quad (4.29)$$

$$|I_L| = \sqrt{3}|I_\phi| = \sqrt{3}|I| \quad (4.30)$$

$$|I_\phi| = |I| = \frac{|I_L|}{\sqrt{3}}$$

$$P_{3\phi} = 3|V||I|\cos\theta = 3|V_\phi||I_\phi|\cos\theta = 3\,|V_{LL}|\frac{|I_L|}{\sqrt{3}}\cos\theta = \sqrt{3}|V_{LL}||I_L|\cos\theta \quad (4.31)$$

The real power equation in terms of line quantities is the same for three-phase wye-connected and delta-connected loads. In the above real power equation, θ is the load phase angle or the angle between load phase voltage and load phase current.

Similarly, we can derive reactive power and apparent power equations in terms of line quantities for three-phase wye-connected and delta-connected loads. We obtain the following relationships for reactive power and apparent power in terms of line quantities for each of the two loads:

$$Q_{3\phi} = \sqrt{3}\,|V_{LL}||I_L|\sin\theta \quad (4.32)$$

$$S_{3\phi} = \sqrt{3}\,|V_{LL}||I_L| \quad (4.33)$$

Example 4.6

A three-phase induction motor with a lagging power factor consumes 12.50 [kW] real power when the line voltage is 480 [V] and the line current is 20 [A]. Determine the apparent power and reactive power supplied and the power factor of the motor.

Solution

The apparent power is:

$$S_{3\phi} = \sqrt{3}|V_{LL}||I_L| = \sqrt{3}\times 480\times 20 = 16627.70\ [VA] \approx 16.63\ [\text{kVA}]$$

The reactive power is:

$$Q_{3\phi} = \sqrt{S_{3\phi}{}^2 - P_{3\phi}{}^2} = \sqrt{16.63^2 - 12.50^2} = 10.96\ [\text{kVAR}]$$

The real power is:

$$P_{3\phi} = S_{3\phi} \times \cos\theta = 12.50\ [\text{kW}]$$

The power factor therefore is:

$$\text{PF} = \cos\theta = \frac{P_{3\phi}}{S_{3\phi}} = \frac{12.50}{16.63} = 0.75 \text{ lagging}$$

Example 4.7

Determine the three-phase real, reactive, and apparent powers supplied to the Δ-connected load in Example 4.4.

Solution

The line voltage magnitude is:

$$|\boldsymbol{V}_{LL}| = |\boldsymbol{V}_{ab}| = |\boldsymbol{V}_{bc}| = |\boldsymbol{V}_{ca}| = 207.84\ [\text{V}]$$

The line current magnitude is:

$$|\boldsymbol{I}_L| = |\boldsymbol{I}_a| = |\boldsymbol{I}_b| = |\boldsymbol{I}_c| = 36\ [\text{A}]$$

The Δ-connected balanced per-phase load impedance is:

$$\boldsymbol{Z}_\Delta = 10\angle 60°\ [\Omega]$$

The load phase angle θ is 60°.
The real power is:

$$P_{3\phi} = \sqrt{3}\,|\boldsymbol{V}_{LL}||\boldsymbol{I}_L|\cos\theta = \sqrt{3} \times 207.84 \times 36 \times \cos 60° = 6480\ [\text{W}]$$
$$= 6.48\ [\text{kW}]$$

The reactive power is:

$$Q_{3\phi} = \sqrt{3}\,|\boldsymbol{V}_{LL}||\boldsymbol{I}_L|\sin\theta = \sqrt{3} \times 207.84 \times 36 \times \sin 60° = 11223\ [\text{VAR}]$$
$$= 11.22\ [\text{kVAR}]$$

The apparent power is:

$$S_{3\phi} = \sqrt{3}\,|\boldsymbol{V}_{LL}||\boldsymbol{I}_L| = \sqrt{3} \times 207.84 \times 36 = 12960\ [\text{VA}] = 12.96\ [\text{kVA}]$$

Example 4.8

A three-phase induction motor is connected to a balanced three-phase voltage source with a line voltage of 480 [V], and it consumes 40 [kW] real power at a lagging power factor of 0.80. Determine the magnitude of the line current.

Solution

The line voltage magnitude is:

$$|\boldsymbol{V}_{LL}| = 480 \ [\mathrm{V}]$$

The real power consumed is:

$$P_{3\phi} = 40 \ [\mathrm{kW}]$$

The induction motor power factor is:

$$\mathrm{PF} = \cos\theta = 0.80 \text{ lagging}$$

We have:

$$P_{3\phi} = \sqrt{3}\,|\boldsymbol{V}_{LL}||\boldsymbol{I}_L|\cos\theta$$

$$40000 = \sqrt{3} \times 480 \times |\boldsymbol{I}_L| \times 0.80$$

$$|\boldsymbol{I}_L| = \frac{40000}{\sqrt{3} \times 480 \times 0.80} = 60.14 \ [\mathrm{A}]$$

The line current magnitude therefore is 60.14 [A].

4.8 ANALYSIS OF BALANCED THREE-PHASE SYSTEMS

The per-phase equivalent circuit analysis method can be used to calculate voltages, currents, and thus powers at different locations in balanced three-phase systems. A wye-connected source supplying power to a wye-connected load through a three-phase transmission line is shown in the first part of Figure 4.12 to illustrate the method. The neutrals of the three-phase source and the three-phase load are isolated. If they are, however, connected as shown by the dotted line, no current will flow in the neutral connection as the three-phase system is balanced and $\boldsymbol{I}_a + \boldsymbol{I}_b + \boldsymbol{I}_c = 0$. The voltages and currents in the three phases of the circuit have equal magnitudes and a progressive phase difference of 120° as described earlier. It is therefore sufficient to analyze the simplified circuit shown in the second part of Figure 4.12, which consists of one phase and the neutral of the original three-phase system. The simplified circuit is solved to determine voltages and currents at various locations in the

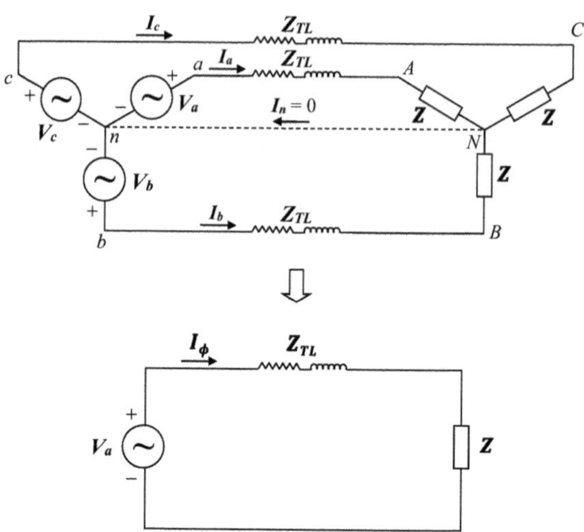

FIGURE 4.12 Three-phase wye-wye circuit connection and its per-phase equivalent circuit.

circuit. The voltages and currents in the other two phases will be equal in magnitude to those obtained for the simplified circuit; however, they will have a progressive phase difference of 120°. If the three-phase circuit contains delta-connected loads, the loads can be easily converted to equivalent wye-connected loads using delta-wye transformations as described earlier, and per-phase analysis can then be performed [1,2,14]. Per-phase equivalent circuit analysis of three-phase systems is demonstrated by means of an example below.

Example 4.9

A 400 [V] three-phase power system is similar to the one shown in the first part of Figure 4.12. It consists of an ideal wye-connected three-phase source with phase *a* voltage of $230.94\angle 0° \ [\text{V}]$ (line voltage 400 [V]) supplying power to a wye-connected load through a three-phase transmission line. The impedances of the transmission line and load are $\boldsymbol{Z}_{TL} = 0.1 + j0.18 \ [\Omega/\text{phase}]$ and $\boldsymbol{Z} = 10 + j15 \ [\Omega/\text{phase}]$, respectively.

For the example three-phase power system, determine:

a. Line current magnitude $|\boldsymbol{I}_L|$ and load phase current magnitude $|\boldsymbol{I}_\phi|$.
b. Magnitudes of load line voltage $|\boldsymbol{V}_{LL,load}|$ and load phase voltage $|\boldsymbol{V}_{\phi,\text{load}}|$.
c. Load real power P_{load}, load reactive power Q_{load}, load apparent power S_{load}, and load power factor PF_{load}.
d. Real power P_{TL}, reactive power Q_{TL}, and apparent power S_{TL} consumed by the transmission line.
e. Source real power P_S, source reactive power Q_S, source apparent power S_S, and source power factor PF_S.

Solution

The voltage source and load in this example three-phase power system are wye-connected, and therefore, the resulting per-phase equivalent circuit is similar to the second part of Figure 4.12.

a. $$\boldsymbol{I}_a = \boldsymbol{I}_\phi = \boldsymbol{I}_L = \frac{\boldsymbol{V}_a}{[\boldsymbol{Z}_{TL} + \boldsymbol{Z}]} = \frac{230.94\angle 0^\circ}{(0.1 + j0.18) + (10 + j15)} = 12.67\angle -56.36^\circ \text{ [A]}$$

$$|\boldsymbol{I}_L| = |\boldsymbol{I}_\phi| = |\boldsymbol{I}_a| = 12.67 \text{ [A]}$$

b. $$\boldsymbol{V}_{\phi,\text{load}} = \boldsymbol{I}_\phi \times \boldsymbol{Z} = \boldsymbol{I}_a \times \boldsymbol{Z} = [12.67\angle -56.36^\circ] \times [10 + j15]$$

$$= [12.67\angle -56.36^\circ] \times [18.03\angle 56.31^\circ]$$

$$= 228.34\angle -0.05^\circ \text{ [V]}$$

$$|\boldsymbol{V}_{\phi,\text{load}}| = 228.34 \text{ [V]}$$

$$|\boldsymbol{V}_{LL,\text{load}}| = \sqrt{3}\, |\boldsymbol{V}_{\phi,\text{load}}| = 395.50 \text{ [V]}$$

c. $$P_{\text{load}} = 3\, |\boldsymbol{V}_{\phi,\text{load}}|\, |\boldsymbol{I}_\phi| \cos\theta = 3 \times 228.34 \times 12.67 \times \cos 56.31^\circ$$

$$= 4814.35 \text{ [W]}$$

$$Q_{\text{load}} = 3\, |\boldsymbol{V}_{\phi,\text{load}}|\, |\boldsymbol{I}_\phi| \sin\theta = 3 \times 228.34 \times 12.67 \times \sin 56.31^\circ$$

$$= 7221.54 \text{ [VAR]}$$

$$S_{\text{load}} = 3\, |\boldsymbol{V}_{\phi,\text{load}}|\, |\boldsymbol{I}_\phi| = 3 \times 228.34 \times 12.67 = 8679.20 \, [VA]$$

$$\text{PF}_{\text{load}} = \cos\theta = \cos 56.31^\circ = 0.55 \text{ lagging}$$

d. $$\boldsymbol{Z}_{TL} = 0.1 + j0.18 \, [\Omega/\text{phase}] = 0.206\angle 60.95^\circ \, [\Omega/\text{phase}]$$

$$\boldsymbol{I}_L = 12.67\angle -56.36^\circ \text{ [A]}$$

$$P_{TL} = 3\, |\boldsymbol{I}_L|^2\, |\boldsymbol{Z}_{TL}| \cos\theta_{TL} = 3 \times 12.67^2 \times 0.206 \times \cos 60.95^\circ = 48.17 \text{ [W]}$$

$$Q_{TL} = 3|\boldsymbol{I}_L|^2|\boldsymbol{Z}_{TL}|\sin\theta_{TL} = 3\times12.67^2\times0.206\times\sin60.95° = 86.73\ [\text{VAR}]$$

$$S_{TL} = 3|\boldsymbol{I}_L|^2|\boldsymbol{Z}_{TL}| = 3\times12.67^2\times0.206 = 99.21\ [\text{VA}]$$

e.

$$P_S = P_{TL} + P_{\text{load}} = 48.17 + 4814.35 = 4862.52\ [\text{W}]$$

$$Q_S = Q_{TL} + Q_{\text{load}} = 86.73 + 7221.54 = 7308.27\ [\text{VAR}]$$

$$S_S = \sqrt{P_S^2 + Q_S^2} = \sqrt{4862.52^2 + 7308.27^2} = 8778.09\ [\text{VA}]$$

$$\theta_S = \tan^{-1}\left(\frac{Q_S}{P_S}\right) = \tan^{-1}\left(\frac{7308.27}{4862.52}\right) = 56.36°$$

$$\text{PF}_S = \cos\theta_S = \cos 56.36° = 0.55 \text{ lagging}$$

4.9 POWER FACTOR CORRECTION

The apparent power S consumed by a load or a combination of several loads consists of the real power P and the reactive power Q as described above. The real power component P does the useful work in the load, whereas the reactive power component Q simply keeps circulating between the source and the load without performing any useful work. Consequently, the real power component P should be as large as possible and the reactive power component Q should be as small as possible in the apparent power S supplied to a load.

The ratio of real power P and apparent power S supplied to a load is the load's power factor (PF). If the load has a high PF ($\cos\theta$), the real power component P supplied to the load will be large and the reactive power component Q will be small. If the load PF is unity, which means $\cos\theta = 1$ and $\theta = 0°$, all apparent power supplied to the load is in the form of real power P and the reactive power Q supplied to the load is zero. The higher the PF of the load, the higher the ability of the load to utilize the apparent power supplied to it [1,12].

The line current supplied to a load in the case of a three-phase system is determined from the following relationship:

$$|\boldsymbol{I}_L| = \frac{P}{\sqrt{3}|\boldsymbol{V}_{LL}|\cos\theta} \tag{4.34}$$

For fixed values of load real power (P) and load line-to-line voltage magnitude ($|\boldsymbol{V}_{LL}|$), the load current magnitude ($|\boldsymbol{I}_L|$) is inversely proportional to the load PF ($\cos\theta$). The higher the load PF, the lower the load current, and conversely, the lower the load

PF, the higher the load current. Increased load current resulting from low load PF leads to higher kVA ratings of the equipment for the same real power output, larger conductor sizes to supply the increased load current, increased copper losses, higher voltage drops in the conductors due to increased current, and decreased utilization of the power system's installed capacity.

The low PF of the induction motor loads, which form a large part of the total industrial loads, is a main reason for the low PF of the collection of loads on the power supply system. Other industrial loads such as lighting and heating loads also have low lagging PFs. During the night, the system PF is further reduced since loads connected to the power supply system are lower, which results in higher power supply system voltages and increased transformer magnetizing currents. The low PF of the collection of loads connected to the power supply system affects the economic operation of the power system since the power system capacity is not utilized to its full extent due to the supply of a large amount of reactive power to the inductive loads.

Capacitors are installed as required at various locations in the power system to improve the system PF. Capacitors draw the leading current from the power system, and they thus supply reactive power to it. The leading current drawn, and thus reactive power supplied by capacitors, neutralizes the lagging current drawn by inductive loads and reactive power consumed by them. Consequently, the PF of the power system is increased. An example below demonstrates the use of capacitors for PF correction in power systems.

Example 4.10

A three-phase wye-connected load operates at a line voltage of 1000 [V] and consumes 100 kW at 0.85 lagging PF. To increase overall system PF to 0.95 lagging, delta-connected capacitors are installed. Determine the three-phase kVAR rating of the delta-connected PF correction capacitors and the value of the capacitor connected in each phase. Determine also the line current before and after installation of capacitors for PF correction.

Solution

Load PF is:

$$\text{PF} = \cos\theta = 0.85 \ \text{lagging}$$

$$\theta = \cos^{-1}(0.85) = 31.79°$$

Reactive power consumed by load is:

$$Q = 100\tan 31.79° = 61.98\ [\text{kVAR}]$$

Power factor PF_{new} after installation of the PF correction capacitors is:

$$\text{PF}_{\text{new}} = \cos\theta_{\text{new}} = 0.95 \ \text{lagging}$$

$$\theta_{\text{new}} = \cos^{-1}(0.95) = 18.19°$$

New system reactive power consumption is:

$$Q_{\text{new}} = 100\tan 18.19° = 32.86\ [\text{kVAR}]$$

Reactive power supplied by PF correction capacitors is:

$$Q_C = Q - Q_{\text{new}} = 61.98 - 32.86 = 29.12\ [\text{kVAR}]$$

Phase current (I_C) in delta-connected PF correction capacitors is:

$$3|\boldsymbol{V_{LL}}||\boldsymbol{I_C}| = Q_C$$

$$|\boldsymbol{I_C}| = \frac{Q_C}{3|\boldsymbol{V_{LL}}|} = \frac{29.12\times 10^3}{3\times 1000} = 9.71\ [\text{A}]$$

Reactance (X_C) of the capacitor in each phase is:

$$X_C = \frac{|\boldsymbol{V_{LL}}|}{|\boldsymbol{I_C}|} = \frac{1000}{9.71} = 102.99\ [\Omega]$$

Capacitance (C) in each phase of delta-connected PF correction capacitors is:

$$C = \frac{1}{2\pi f X_C} = \frac{1}{2\pi\times 60\times 102.99} = 15.52\times 10^{-6}[\text{F}] = 25.76\ [\mu\text{F}]$$

Figure 4.13 shows the power triangle of Example 4.10.

Line current drawn by the load before PF correction capacitor installation is:

$$|\boldsymbol{I_L}| = \frac{P}{\sqrt{3}|\boldsymbol{V_{LL}}|\cos\theta} = \frac{100000}{\sqrt{3}\times 1000\times 0.85} = 67.92\ [\text{A}]$$

Line current drawn by the load after capacitor bank installation is:

$$|\boldsymbol{I_{L\,\text{new}}}| = \frac{P}{\sqrt{3}|\boldsymbol{V_{LL}}|\cos\theta_{\text{new}}} = \frac{100000}{\sqrt{3}\times 1000\times 0.95} = 60.77\ [\text{A}]$$

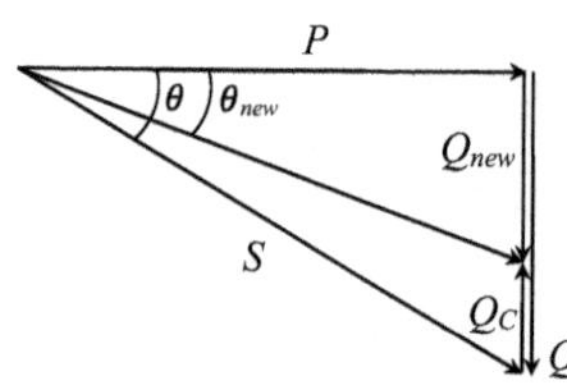

FIGURE 4.13 Power triangle of Example 4.10.

4.10 SUMMARY

The electric power generated in power plants is of three-phase AC form. Three-phase AC power systems are preferred over single-phase AC power systems due to their many advantages such as the possibility to transmit larger amounts of power for similar conductor dimensions, constancy of power supplied in three-phase systems, and higher power output of three-phase machines than single-phase machines of similar size.

A three-phase voltage source consists of three single-phase voltage sources which have equal voltage magnitude and a progressive phase difference of 120°. The three-phase power system operation is always balanced with balanced three-phase voltage sources and equal load impedances in the three phases.

The three-phase voltage sources are mainly connected in wye configuration, whereas the three-phase loads can be connected in wye or delta configuration. The power supplied to a balanced three-phase load is constant at all times, and it does not pulsate with time as the power supplied to a single-phase load.

The per-phase equivalent circuit method is used for the analysis of three-phase power systems, which allows the determination of the power system voltages and currents at various locations in one of the phases. Voltages and currents in the other two phases of the power system will have identical magnitude and a progressive phase difference of 120°.

The loads connected to the power system should have high PFs preferably close to unity for higher use of power supplied and improved utilization of the installed capacity of the power system. PF correction capacitors are installed at various locations in the power network to improve the system PF. They draw the leading current and supply reactive power to the power system, and they neutralize the lagging current drawn by inductive loads and reactive power consumed by them. The power factor of the power system is consequently improved.

REVIEW QUESTIONS

4.1 What are the reasons for the generation of three-phase AC power in power plants?

4.2 What is the meaning of the term balanced in a balanced three-phase system?

4.3 Describe why the current in the neutral wire of a balanced three-phase system is zero.

4.4 What does the term phase sequence mean in a three-phase system?

4.5 Describe why three-phase voltage sources are generally not connected in delta configuration.

4.6 Derive the relationships for phase and line voltages and currents for a balanced wye-connected source supplying power to a balanced wye-connected load.

4.7 Derive the relationships for phase and line voltages and currents for a balanced wye-connected source supplying power to a balanced delta-connected load.

4.8 Describe how a delta-connected load can be converted to a wye-connected load using the delta-wye conversion.
4.9 Derive the relationships for real power, reactive power, and apparent power in a three-phase AC circuit.
4.10 Describe how the total power supplied to a balanced three-phase load in a balanced three-phase power system is constant at all times.
4.11 Describe how a three-phase system is analyzed by constructing its per-phase equivalent circuit.
4.12 Describe the procedure for the correction of low power factor in a power system.

5 Power System Components Modeling

5.1 INTRODUCTION

In this chapter, we study various components of an electric power system such as synchronous generators, transformers, transmission lines, and power system loads, and we shall develop their equivalent circuit models. These component models are subsequently used to build models of power systems, which are interconnections of various power system components.

The synchronous generator is driven by a prime mover such as a steam turbine, and it converts mechanical energy to electrical energy in a three-phase alternating current form. The transformer is a device that is used to transform power system voltage from one level to another. It serves as the link between the generator and the transmission line and between the transmission line and the distribution system, which supplies electrical energy through still other transformers to the loads on the power system.

This chapter is divided into four parts with each part describing each of the main electric power system components, namely synchronous generators, transformers, transmission lines, and power system loads.

Part I: Synchronous Generators

5.2 SYNCHRONOUS GENERATORS: INTRODUCTION

Synchronous generators are machines which transform mechanical energy supplied using a prime mover such as a steam turbine into AC electric power. The two main components of a synchronous generator are the rotor and the stator as shown in Figure 5.1. DC is supplied to the field winding of the rotor, and the rotor is rotated using a prime mover. It results in a rotating magnetic field within the synchronous machine. The rotating magnetic field induces three-phase voltages in the three-phase stator windings $a-a'$, $b-b'$, and $c-c'$ [7,26,45].

The relationship between electrical frequency induced in stator windings and rotor magnetic field rotation speed is as follows:

$$f_e = \frac{n_m P}{120} \tag{5.1}$$

DOI: 10.1201/9781003432340-5

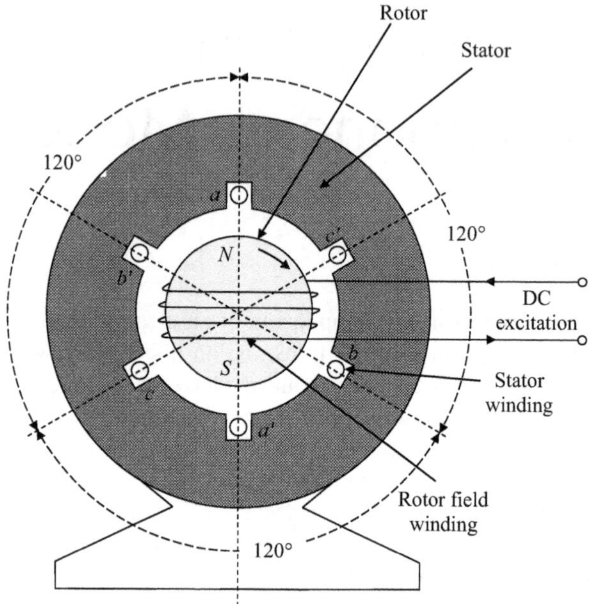

FIGURE 5.1 Schematic diagram of a two-pole round-rotor synchronous generator [II].

where

f_e = stator electrical frequency [Hz]
n_m = mechanical speed of rotor magnetic field or rotor [rpm]
P = number of poles

Electric power is generated at 60 Hz frequency in the USA and at 50 Hz frequency in Europe. From the above equation, to generate 60-Hz power frequency in a two-pole synchronous machine, the rotor must turn at 3600 [rpm]. Similarly, to generate 50-Hz frequency power in a four-pole machine, the rotor must turn at 1500 [rpm]. Windings of synchronous machines with four or more poles are created by repeating the three-phase winding pattern of simple two-pole machines. The reader is requested to refer to any standard AC electrical machine text to learn about the construction of synchronous machines in detail.

The magnitude of the voltage $\boldsymbol{E}$ induced in any stator phase winding is determined from the following equation:

$$|\boldsymbol{E}| = K|\phi_f|\omega \tag{5.2}$$

K is a constant dependent upon the construction of the machine, ϕ_f is the flux produced by the rotor in the synchronous machine in Weber, and ω is the angular frequency of rotation of the rotor in radians per second [7].

5.3 EQUIVALENT CIRCUIT OF A SYNCHRONOUS GENERATOR

The internal generated voltage in phase a of a synchronous generator is $\boldsymbol{E}\ (=\boldsymbol{E}_a)$. This voltage is, however, not generally equal to the generator terminal voltage $\boldsymbol{V}_t\ (=\boldsymbol{V}_a)$. Only when the synchronous generator is unloaded and there is no current flowing in the stator armature windings, the internal generated voltage $\boldsymbol{E}$ equals the terminal voltage $\boldsymbol{V}_t$. The internal voltage differs from the terminal voltage for loaded generators due to distortion of rotor air gap magnetic field produced by stator winding current, leakage reactance of stator windings, and resistance of stator windings [7,30,38].

The most prominent reason for the difference between $\boldsymbol{E}$ and $\boldsymbol{V}_t$ is distortion of rotor air gap magnetic field by stator winding current. Rotation of the synchronous generator rotor results in induction of voltage $\boldsymbol{E}$ in the generator's stator windings. If the synchronous generator is loaded, current flows through the stator armature windings. Three-phase currents flowing through the stator windings, however, produce a rotating magnetic field of their own in the synchronous machine. Due to the presence of a stator magnetic field, the original rotor magnetic field becomes distorted, resulting in an alteration in phase voltage. This effect is called the armature reaction.

The rotor magnetic field $\boldsymbol{B}_{\text{field}}$ produces an internal generated voltage $\boldsymbol{E}$ in phase a whose peak value coincides with the direction of $\boldsymbol{B}_{\text{field}}$. On no load, there is no armature current flow, and $\boldsymbol{E}$ will be equal to the terminal voltage $\boldsymbol{V}_t$. As a balanced steady load current is drawn from the three-phase stator winding, the stator currents of phases a, b, and c together produce a synchronously rotating field $\boldsymbol{B}_s$ in the direction of rotation of the rotor. This rotating field produced by three-phase stator currents, called armature reaction magnetic field or stator magnetic field, is stationary with respect to the rotor magnetic field $\boldsymbol{B}_{\text{field}}$ which rotates with the rotor field windings. The resultant air gap magnetic field $\boldsymbol{B}_r$ is the phasor sum of the rotor magnetic field and the stator magnetic field:

$$\boldsymbol{B}_r = \boldsymbol{B}_{\text{field}} + \boldsymbol{B}_s \tag{5.3}$$

The stator magnetic field $\boldsymbol{B}_s$ produces a voltage of its own in the stator, and this voltage is called $\boldsymbol{E}_s$.

With two voltages present in the stator windings, the terminal voltage is the sum of the internal generated voltage $\boldsymbol{E}$ and the armature reaction voltage $\boldsymbol{E}_s$:

$$\boldsymbol{V}_t = \boldsymbol{E} + \boldsymbol{E}_s \tag{5.4}$$

The armature reaction voltage $\boldsymbol{E}_s$ in phase a lags 90° behind phase $\boldsymbol{a}$ current $\boldsymbol{I}_a$. The voltage $\boldsymbol{E}_s$ is also directly proportional to the current $\boldsymbol{I}_a$. If X_{ar} is a constant of proportionality, then the armature reaction voltage can be expressed as:

$$\boldsymbol{E}_s = -jX_{\text{ar}}\boldsymbol{I}_a \tag{5.5}$$

We therefore obtain the following expression for $\boldsymbol{V}_t$:

$$\boldsymbol{V}_t = \boldsymbol{E} - jX_{\text{ar}}\boldsymbol{I}_a \tag{5.6}$$

X_{ar} in the above equation is the inductive reactance, which accounts for the effect of the armature reaction. After including the effects of stator winding leakage reactance X_l and stator winding resistance R_a, we obtain the complete circuit model of the synchronous generator. We thus have:

$$\boldsymbol{V_t} = \boldsymbol{E} - jX_{ar}\boldsymbol{I_a} - jX_l\boldsymbol{I_a} - \boldsymbol{R_a I_a} \tag{5.7}$$

The reactances X_{ar} and X_l in the above equation are combined, and the total reactance $(X_{ar} + X_l) = X_s$ is called synchronous reactance of the machine. We therefore rewrite equation (5.7) as follows:

$$\boldsymbol{V_t} = \boldsymbol{E} - jX_s\boldsymbol{I_a} - \boldsymbol{R_a I_a} \tag{5.8}$$

The circuit model of a synchronous generator is shown in Figure 5.2. Armature resistance R_a value is less than 1% of synchronous reactance X_s value, and it is generally neglected in power system analysis. Therefore, the simplified circuit model of Figure 5.3 is generally used instead of the circuit model of Figure 5.2. We therefore have for the circuit in Figure 5.3:

$$\boldsymbol{V_t} = \boldsymbol{E} - jX_s\boldsymbol{I_a} \tag{5.9}$$

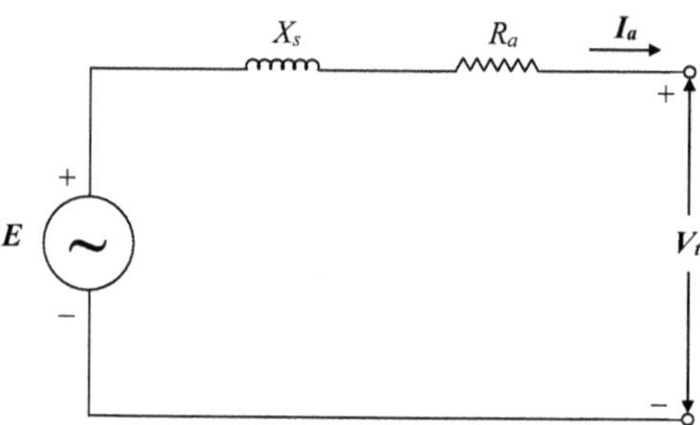

FIGURE 5.2 Equivalent circuit model of a synchronous generator.

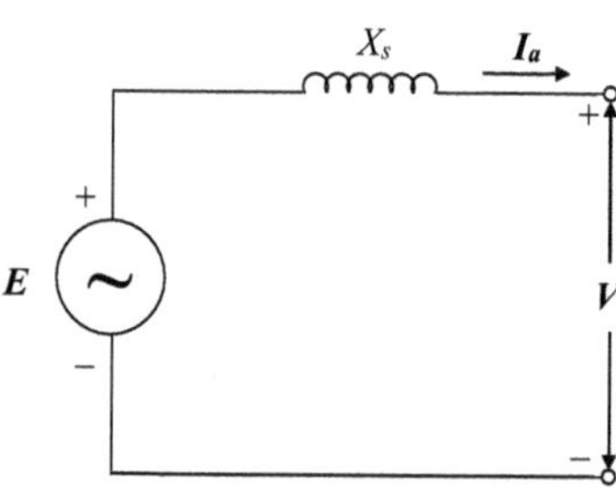

FIGURE 5.3 Simplified circuit model of a synchronous generator.

5.4 SYNCHRONIZATION OF THE SYNCHRONOUS GENERATOR TO THE GRID

To connect a synchronous generator to the power grid, certain requirements must be met. The frequency of the grid must be equal to the generator emf frequency. The phase sequence *abc* or *acb* of the grid must be similar to that of the synchronous generator. The magnitude of the generator emf is controlled by varying the rotor field current, and it must be equal to the magnitude of grid voltage. The phase difference between the synchronous generator emf and the grid voltage must also be zero. For phase difference control, the speed of the synchronous generator prime mover is regulated extremely gently until phase angles of the three-phase voltages of the synchronous generator and grid are equal. If all these requirements are met, a synchronous generator can be connected to the power grid to supply power [12].

A simplified circuit for synchronizing the synchronous machine to the power grid is shown in Figure 5.4. The three lamps shown blink 120 times each second due to 60 Hz grid frequency when switches *a*, *b*, and *c* are kept open and the synchronous machine is at a standstill and not generating any power. It is, however, not possible to view the blinking directly due to its high frequency.

Using a prime mover, torque is applied to the synchronous generator, and as the machine rotates, the lamps blink more slowly at a rate of twice the difference between synchronous machine emf frequency and grid frequency. When all requirements for synchronization described above are met, the lamps are turned off simultaneously. The interconnected switches *a*, *b*, and *c* are subsequently closed, and the synchronous machine is then synchronized to the power grid.

The frequency of the synchronous generator is kept slightly higher than the frequency of the power grid, since it will result in a synchronous generator supplying

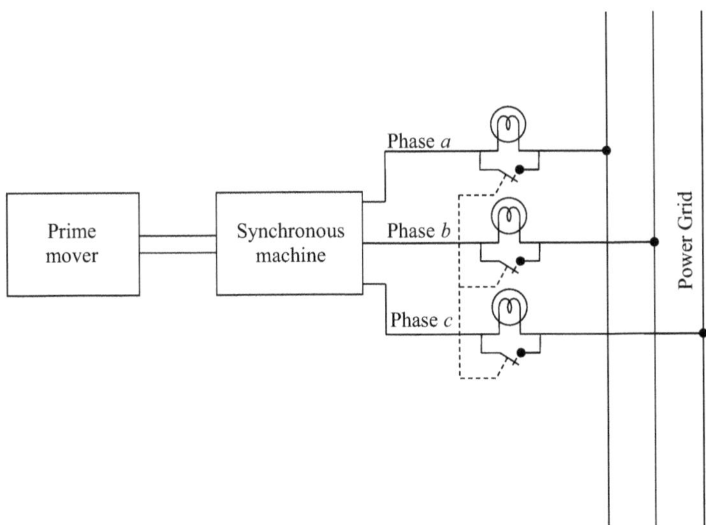

FIGURE 5.4 Simplified circuit for synchronization of the synchronous machine to the power grid [12].

power to the power grid at the time of synchronization rather than drawing power from it as a synchronous motor.

In large synchronous generators, the complete process of synchronizing a new generator to the power grid is automated using computers. In the case of smaller generators, the synchronization work is performed manually as described above.

5.5 PHASOR DIAGRAMS OF A SYNCHRONOUS GENERATOR SUPPLYING POWER TO LOADS WITH DIFFERENT POWER FACTORS

The AC voltages generated in synchronous generators are expressed as phasors as described earlier [7,14]. Since each phasor has magnitude and angle, it is possible to express the relationships between different phasors using a two-dimensional diagram. Such a diagram is called a phasor diagram [7].

The phasor diagrams of a synchronous generator supplying power to loads having unity power factor, lagging power factor, and leading power factor are shown in Figure 5.5. The terminal voltage $\boldsymbol{V_t}$ is assumed to be the reference phasor having a phase angle of 0° for all power factor loads. As described earlier, $\boldsymbol{V_t}$ for all power factor loads is obtained by subtracting voltage drops in armature resistance R_a and

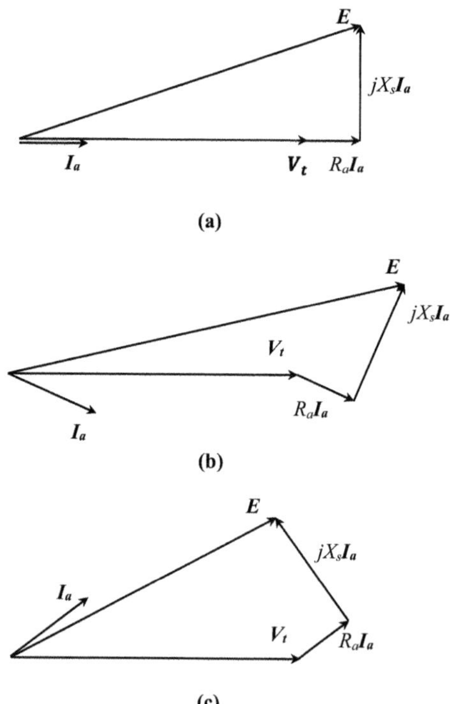

FIGURE 5.5 Phasor diagram of a synchronous generator supplying power to loads having unity power factor, lagging power factor, and leading power factor [14].

synchronous reactance X_s due to the flow of phase current $\boldsymbol{I}_a$ from the total generated phase voltage $\boldsymbol{E}$.

Figure 5.5 shows that for any specified terminal voltage $\boldsymbol{V}_t$ and magnitude of phase current $\boldsymbol{I}_a$, a larger value of total generated phase voltage $\boldsymbol{E}$ is required for lagging power factor loads than for leading power factor loads. The generated phase voltage $\boldsymbol{E}$ for unity power factor load is between the generated phase voltages for lagging power factor and leading power factor loads. Since the generated phase voltage $\boldsymbol{E}$ is directly proportional to the field current assuming linearity, to maintain equal terminal voltage, it is required to supply a larger field current for lagging power factor loads than for leading power factor loads. The frequency of the system is maintained constant by keeping ω constant.

Using synchronous machine field current control, it is therefore possible to control the flow of reactive power. In the second part of Figure 5.5, the synchronous generator is over-excited and it supplies lagging current and reactive power to the system similar to a capacitor. In the third part of Figure 5.5, the synchronous generator is under-excited and it supplies leading current to the system and draws reactive power similar to an inductor [7,14].

As mentioned earlier, in real synchronous machines, the synchronous reactance X_s is much larger than the armature resistance R_a, and therefore, R_a is many times neglected in the power system analysis.

Example 5.1

A 1000 [V] Y-connected three-phase synchronous generator has a per-phase synchronous reactance of 2 [Ω], and the armature resistance is ignored. The field current is regulated to have generated line-to-line voltage of 1000 [V] at no load. Determine the terminal voltage of the generator if it is connected to a load to supply the rated current of 100 [A] at 0.95 PF lagging.

Solution

The generator is Y-connected, and therefore, the generated phase voltage is:

$$|\boldsymbol{E}| = 1000/\sqrt{3} = 577.35\ [\text{V}].$$

At no load, the armature current $\boldsymbol{I}_a$ is zero, and therefore, $jX_s\boldsymbol{I}_a = 0$.

$R_a\boldsymbol{I}_a$ voltage drop is also zero since $R_a = 0$.

We therefore have:

$$|\boldsymbol{V}_t| = |\boldsymbol{E}| = 577.35\ [\text{V}].$$

If the generator is connected to a load to supply the rated current of 100 [A] at 0.95 PF lagging, we have:

$$\boldsymbol{I}_a = 100\angle -18.2°\ [\text{A}].$$

The resulting phasor diagram is shown in Figure 5.6.

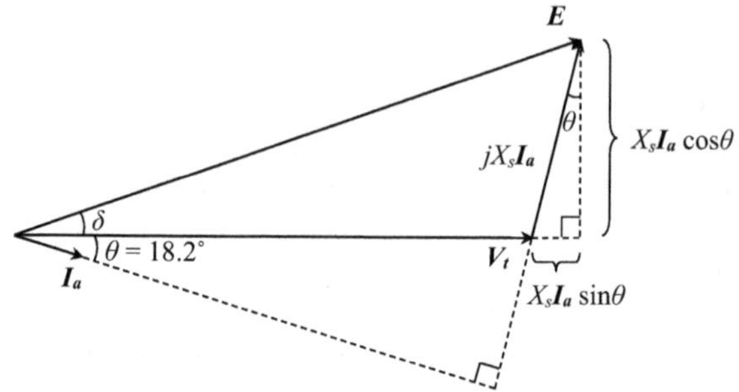

FIGURE 5.6 Synchronous generator phasor diagram for the example.

In the phasor diagram shown, $\boldsymbol{v}_t$ is assumed to be the reference phasor having a phase angle of 0°.

The magnitude of $\boldsymbol{E}$ is 577.35 [V], and the quantity $jX_s\boldsymbol{I}_a$ is:

$$jX_s\boldsymbol{I}_a = j(2.0)\left(100\angle -\cos^{-1}(0.95)\right) = j(2.0)(100\angle -18.2°) = 200\angle 71.8° \text{ [V]}$$

The unknowns in the phasor diagram are the magnitude of $\boldsymbol{V}_t$ and the angle δ by which $\boldsymbol{E}$ leads $\boldsymbol{V}_t$.

From Figure 5.6, we have:

$$|\boldsymbol{E}|^2 = (|\boldsymbol{V}_t| + X_s\,|\,\boldsymbol{I}_a\,|\sin\theta)^2 + (X_s\,|\,\boldsymbol{I}_a\,|\cos\theta)^2$$

The terminal phase voltage at the rated current of 100 [A] at 0.95 PF lagging supplied to the load is calculated as follows:

$$(577.35)^2 = \left[|\boldsymbol{V}_t| + (2.0)(100)\sin 18.2°\right]^2 + \left((2.0)(100)\cos 18.2°\right)^2$$

$$333333 = \left[|\boldsymbol{V}_t| + 62.47\right]^2 + 36098$$

$$297235 = \left[|\boldsymbol{V}_t| + 62.47\right]^2$$

$$545.19 = |\boldsymbol{V}_t| + 62.47$$

$$|\boldsymbol{V}_t| = 482.72 \text{ [V]}$$

Since the generator is Y-connected, terminal line-to-line voltage $= \sqrt{3}|\boldsymbol{V}_t| = 836.10$ [V].

Part II: Power Transformers

5.6 IDEAL TRANSFORMER

An ideal transformer as shown in Figure 5.7 is a lossless electromagnetic device which has an input winding and an output winding coupled through a mutual flux. Simple equations described below provide the relationships between voltages of input and output windings and between currents in input and output windings.

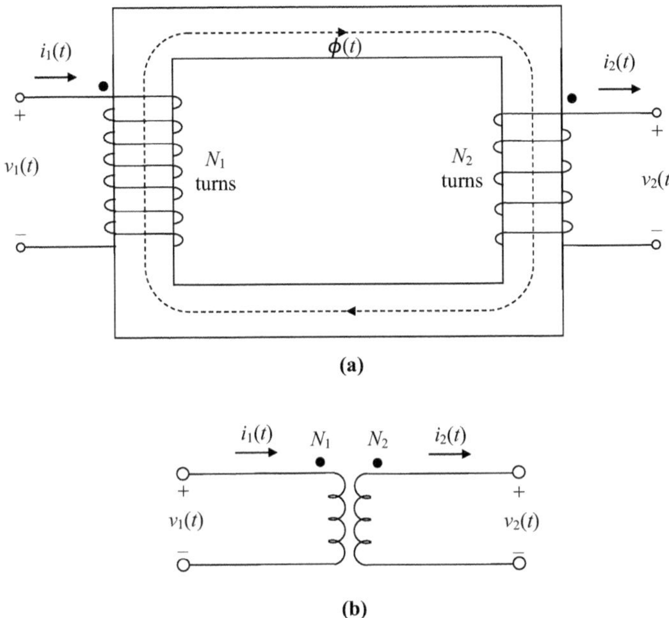

FIGURE 5.7 (a) Ideal transformer and (b) ideal transformer symbol [12].

The transformer shown above has N_1 turns in the primary winding and N_2 turns in the secondary winding. The following equation describes the relationship between voltages $v_1(t)$ and $v_2(t)$, respectively, on the primary side and the secondary side of the transformer:

$$\frac{v_1(t)}{v_2(t)} = \frac{N_1}{N_2} = a \tag{5.10}$$

In the above equation, a is the turns ratio of the transformer.

$$a = \frac{N_1}{N_2} \tag{5.11}$$

The following equation describes the relationship between currents $i_1(t)$ and $i_2(t)$, respectively, flowing into the primary winding and coming out of the secondary winding of the transformer:

$$N_1 i_1(t) = N_2 i_2(t)$$

$$\frac{i_1(t)}{i_2(t)} = \frac{N_2}{N_1} = \frac{1}{a} \tag{5.12}$$

In phasor form, we can write the above equations as follows:

$$\frac{V_1}{V_2} = a \tag{5.13}$$

$$\frac{I_1}{I_2} = \frac{1}{a} \tag{5.14}$$

The phase angles of input voltage $v_1(t)$ and output voltage $v_2(t)$ are equal. Similarly, the phase angles of input current $i_1(t)$ and output current $i_2(t)$ are also equal. The phase angles of input side and output side voltages and currents therefore remain untransformed; however, their magnitudes are transformed according to the turns ratio.

Dot convention is utilized in transformers to describe the connection between voltage polarities of the primary winding and the secondary winding at any instant. If the primary winding voltage has positive polarity at the dotted end with respect to the undotted end, then the secondary winding voltage will also have positive polarity at the dotted end with respect to the undotted end [7,8,12,32]. If the primary side current enters the primary winding at the dotted end, the secondary side current will come out of the secondary winding at the dotted end.

We next determine the relationships between power entering and exiting an ideal transformer. The equation for the power entering the primary side of the transformer is:

$$P_{\text{in}} = |V_1||I_1|\cos\theta_1 \tag{5.15}$$

θ_1 in the above equation is the angle between the primary side voltage and the primary side current.

The equation for the power exiting the secondary side of the transformer is:

$$P_{\text{out}} = |V_2||I_2|\cos\theta_2 \tag{5.16}$$

θ_2 in the above equation is the angle between the secondary side voltage and the secondary side current.

Since the phase angles of input side and output side voltages and currents remain untransformed, we have:

$$\theta_1 = \theta_2 = \theta \tag{5.17}$$

The power factor on the primary side of the transformer is therefore the same as that on the secondary side of the transformer.

Using equations (5.13), (5.14), and (5.17) on equation (5.16), we further obtain:

$$P_{\text{out}} = |V_2||I_2|\cos\theta_2 = |V_2||I_2|\cos\theta = \frac{|V_1|}{a}\left(a|I_1|\right)\cos\theta = |V_1||I_1|\cos\theta = P_{\text{in}} \tag{5.18}$$

The input power and output power for an ideal transformer are therefore equal [7,12].

The relationships obtained above for real power are also applicable to reactive power and apparent power. We can therefore write:

$$Q_{in} = |V_1||I_1|\sin\theta = |V_2||I_2|\sin\theta = Q_{out} \tag{5.19}$$

$$S_{in} = |V_1||I_1| = |V_2||I_2| = S_{out} \tag{5.20}$$

We next determine how any impedance is transformed as it is referred from the secondary side to the primary side of the transformer and vice versa. The impedance of any circuit element or load is the ratio of voltage across the circuit element and the current flowing through it.

$$Z_L = \frac{V_L}{I_L} \tag{5.21}$$

If voltage on the secondary side of the transformer is V_2 and current on the secondary side is I_2, the secondary side load impedance is:

$$Z_L = \frac{V_2}{I_2} \tag{5.22}$$

Since the transformer transforms voltages and currents on the primary and secondary sides, it also transforms their ratio and the apparent impedance of the load. The apparent impedance of the secondary side load on the primary side of the transformer is:

$$Z_L' = \frac{V_1}{I_1} \tag{5.23}$$

Using equations (5.13) and (5.14) on the above equation, we obtain:

$$Z_L' = \frac{V_1}{I_1} = \frac{aV_2}{I_2 / a} = a^2 \frac{V_2}{I_2}$$

$$Z_L' = a^2 Z_L \tag{5.24}$$

The impedance Z_L connected across the secondary winding therefore has the same effect as the equivalent impedance $Z_L' = a^2 Z_L$ connected across the primary winding of the transformer. Z_L' is called the secondary impedance referred to the primary side of the transformer [7,12].

5.7 TRANSFORMER LOSSES AND EQUIVALENT CIRCUIT OF REAL NON-IDEAL TRANSFORMER

In the above description, an ideal transformer has been considered, which is assumed to have no losses. A real non-ideal transformer differs from an ideal transformer since the real transformer has resistive losses in its primary and secondary windings and core losses comprising hysteresis and eddy current losses. Also, a real non-ideal transformer's core is not perfectly permeable, and a finite amount of magnetizing current is required to produce magnetic flux inside a real ferromagnetic core. Furthermore, not all fluxes link with the primary and secondary windings simultaneously due to leakages [7,21,29,36].

Power losses occur due to resistance of the primary and secondary windings of the transformer. To account for these losses, resistances R_1 and R_2, respectively, are placed in the primary circuit and the secondary circuit of the transformer.

When an AC power source is connected to a real transformer, a current flows in the transformer primary circuit, even when the secondary circuit is open-circuited and there is no power supplied to the load. This current is the current required to produce magnetic flux in a real ferromagnetic core. It has two components: magnetization current required to produce flux in the transformer core, and core-loss current required due to hysteresis and eddy current losses.

The core magnetization current is proportional to the voltage applied to the core. It lags the applied voltage by 90°, and therefore, it is modeled using a reactance X_O connected across the primary voltage source. Eddy current losses are resistive heating losses occurring in the core of the transformer. Hysteresis losses occur due to the rearrangement of the magnetic domains in the core during each half-cycle of the AC sinusoidal voltage applied to the primary winding. The core-loss current responsible for eddy current and hysteresis losses is proportional to the voltage applied to the core, and it is also in phase with the applied voltage. It is therefore modeled using a resistance R_O connected across the primary voltage source.

The core flux or mutual flux links both the primary and secondary windings of the transformer. The leakage fluxes of the primary winding and the secondary winding, however, escape the core, and they pass through only one of the transformer windings. These escaped fluxes result in self-inductance in the primary and secondary windings. The leakage fluxes of the primary winding and the secondary winding are modeled using the primary and secondary inductors L_1 and L_2, respectively, with corresponding reactance values of X_1 and X_2, respectively.

When the non-ideal effects of winding resistances, core losses, magnetizing reactance, and leakage reactances are included, the resulting circuit is shown in Figure 5.8, where the primary and the secondary are connected using an ideal transformer. Using voltage, current, and impedance transformation equations described above, the ideal transformer can be removed from Figure 5.8 and the entire equivalent circuit can be referred either to the primary as shown in Figure 5.9a or to the secondary as shown in Figure 5.9b. The reader is requested to refer to any standard electric machinery text to learn about tests used to determine transformer equivalent circuit component values.

The excitation branch consisting of R_O and X_O has much larger impedance, and therefore, the current flowing through it is much less than the load current of the transformer. The excitation branch is sometimes completely deleted in analysis without loss of accuracy [5,7,12,38]. The simplified equivalent circuit of the transformer in such a case is shown in Figure 5.10.

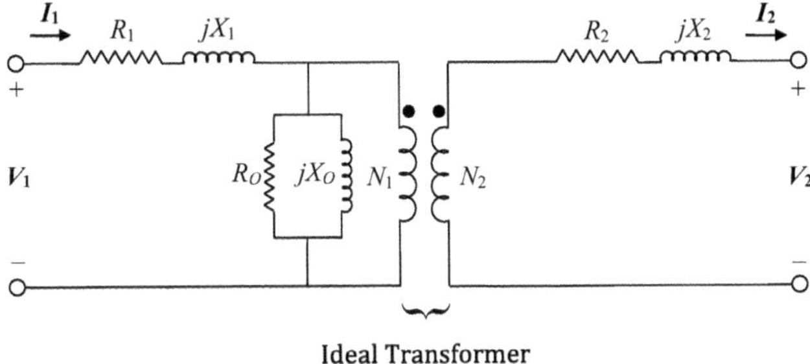

FIGURE 5.8 Model of a real transformer [38].

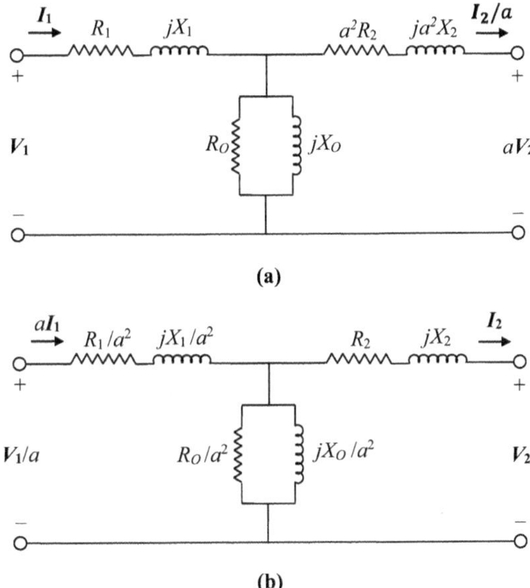

FIGURE 5.9 (a) Transformer model referred to the primary side and (b) transformer model referred to the secondary side [38].

Example 5.2

The ohmic values of the circuit parameters of a transformer having a turns ratio of 10 are $R_1 = 0.8\ [\Omega]$, $R_2 = 0.0082\ [\Omega]$, $X_1 = 4\ [\Omega]$, $X_2 = 0.044\ [\Omega]$, $R_O = 400\ [\Omega]$ referred to the primary, and $X_O = 120\ [\Omega]$ referred to the primary. Determine the transformer equivalent circuit parameters referred to (a) the transformer primary and (b) the transformer secondary.

Solution

a. The transformer equivalent circuit parameters referred to the transformer primary are calculated as follows:

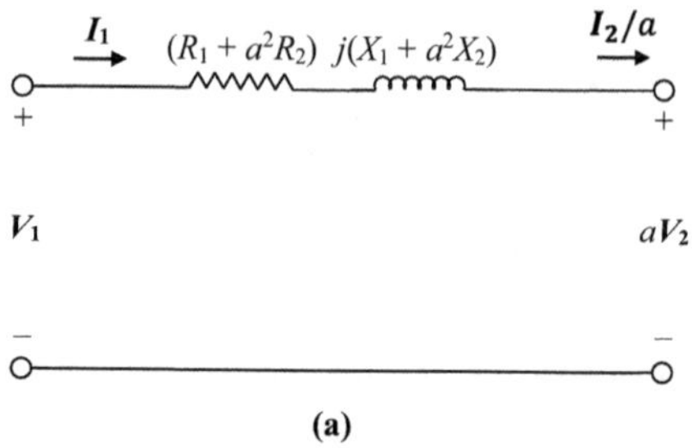

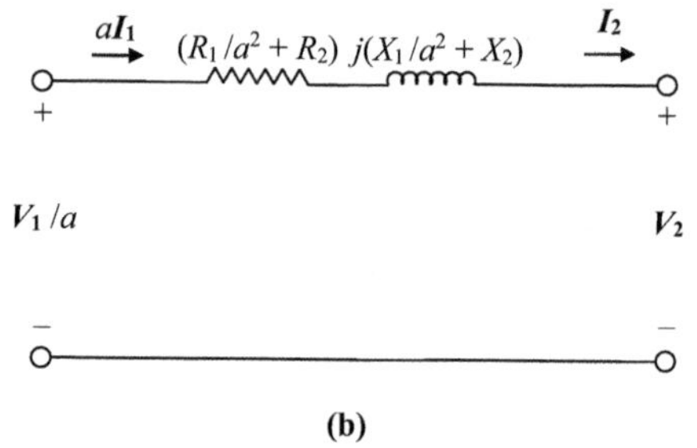

FIGURE 5.10 Approximate transformer models: (a) without excitation branch and referred to the primary side and (b) without excitation branch and referred to the secondary side [14].

$$R_{\text{equivalent1}} = R_1 + a^2 R_2 = 0.8 + (10)^2 (0.0082) = 1.62\ [\Omega]$$

$$X_{\text{equivalent1}} = X_1 + a^2 X_2 = 4 + (10)^2 (0.044) = 8.4\ [\Omega]$$

$$R_{O1} = 400\ [\Omega]$$

$$X_{O1} = 120\ [\Omega]$$

b. The transformer equivalent circuit parameters referred to the transformer secondary are calculated as follows:

$$R_{\text{equivalent2}} = \frac{R_1}{a^2} + R_2 = \frac{0.8}{100} + 0.0082 = 0.0162\ [\Omega]$$

$$X_{\text{equivalent2}} = \frac{X_1}{a^2} + X_2 = \frac{4}{100} + 0.044 = 0.084\ [\Omega]$$

$$R_{O2} = \frac{400}{100} = 4\ [\Omega]$$

$$X_{O2} = \frac{120}{100} = 1.2\ [\Omega]$$

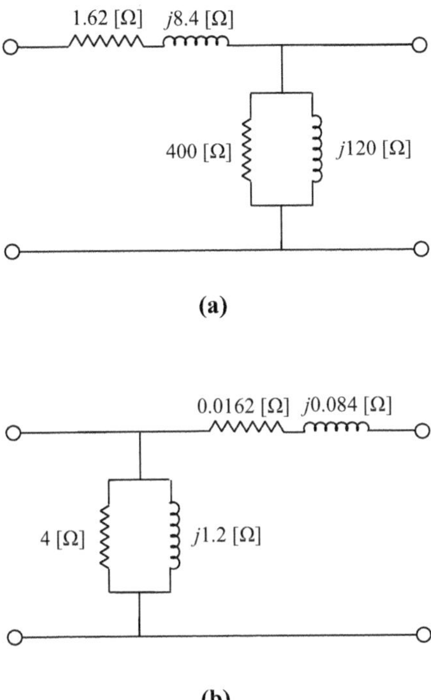

FIGURE 5.11 Transformer equivalent circuit parameters referred to (a) the transformer primary side and (b) the transformer secondary side.

The transformer equivalent circuit parameters referred to the transformer primary side and the transformer secondary side are shown in Figure 5.11a and b.

5.8 THREE-PHASE TRANSFORMERS

A three-phase transformer comprises three transformers, which can be present either as separate transformers or in combined form on the same core [7,38]. Each of the three-phase transformer primary and secondary windings can have either a wye (Y) or delta (Δ) connection. Consequently, four different types of three-phase transformers are possible depending upon the connection of the primary and secondary windings: wye-wye (Y-Y), wye-delta (Y-Δ), delta-wye (Δ-Y) and delta-delta (Δ-Δ).

For the analysis of a three-phase transformer, we consider only one phase of the transformer. Any one phase of a three-phase transformer is exactly similar to a single-phase transformer described above. Three-phase transformer calculations are performed on a per-phase basis using analysis techniques exactly similar to those utilized for single-phase transformers.

Part III: Transmission Lines

5.9 TRANSMISSION LINE PARAMETERS

Electric energy can be transmitted over long distances in an efficient and cost-effective manner using a well-developed, high-capacity network of transmission lines. The main function of the transmission network thus is to transmit electric energy from electric power generating units at various locations to distribution systems, which in turn supply power to the loads. Transmission lines also connect nearby utilities, allowing power transfer within regions under normal circumstances and also power transfers between regions in case of emergency. Most power transmission lines in the world have a three-phase aerial design with bare conductors for transmitting electrical power and surrounding air serving as an insulating medium.

The design parameters of a transmission line for the purpose of power system modeling and analysis are series resistance and inductance, and shunt capacitance and conductance. Using these parameters, it is possible to develop transmission line models which can be utilized subsequently for power system analysis. The transmission line series resistance is dependent on the physical composition of the conductor at a particular temperature. The shunt conductance is there to account for the leakage current flowing between the phase conductors and ground across insulators and ionized pathways in air. The transmission line has inductance due to the effect of magnetic field around the phase conductors, whereas the transmission line capacitance is due to the effect of electric field around the phase conductors.

5.10 TRANSMISSION LINE RESISTANCE AND SHUNT CONDUCTANCE

Even when resistance R and shunt conductance G of a transmission line have a relatively low effect on equivalent transmission line impedances and therefore on transmission capacity, they completely determine the real transmission losses [11,44]. The transmission line conductor DC resistance is obtained from the formula:

$$R_{DC} = \frac{\rho l}{A}\ [\Omega] \tag{5.25}$$

where

ρ = conductor resistivity $[\Omega\ \text{m}]$

l = conductor length [m]

A = conductor cross-sectional area $\left[\text{m}^2\right]$

Due to a phenomenon called the skin effect, the current distribution in a transmission line conductor is not uniform over the conductor cross section during the flow of AC, and the current density is greatest at the conductor's surface. The conductor AC resistance is consequently around 2% higher than DC resistance at 60 [Hz] power frequency. The conductors are also stranded to have flexibility. A stranded conductor is spiraled, and the strands are longer than the finished conductor due to the spiral pattern. The transmission line conductor resistance is therefore higher than the values calculated from the above equation. The conductor resistance also varies linearly with variation in temperature. It is therefore best to determine the transmission line conductor resistance using conductor company data due to the above-mentioned effects.

The shunt conductance G is responsible for the resistive leakage current between the phase conductors and ground across insulation strings and ionized pathways in the air. The leakage current is affected by weather conditions, atmospheric humidity, and pollution. Shunt conductance is neglected during normal power system operating conditions since it is of negligible magnitude.

5.11 TRANSMISSION LINE INDUCTANCE AND CAPACITANCE

The remaining transmission line parameters for the purpose of power system modeling and analysis are transmission line inductance and capacitance [14,48]. The formulas for transmission line inductance and capacitance are mentioned here only to have a simple and more understandable description and to avoid unnecessary derivations. The reader is requested to refer to other power system references for the derivation of formulas for the calculation of these quantities.

The inductance per phase of a single-phase two-wire line is:

$$L = \frac{\mu_0}{2\pi} \ln\left(\frac{d}{re^{-1/4}}\right) \text{ [H / m]} \tag{5.26}$$

We define:

$$r' = re^{-1/4}$$

We also have:

$$\mu_0 = 4\pi \times 10^{-7} \text{ [H / m]}$$

Substituting r' and μ_0 in equation (5.26), we obtain:

$$L = 2 \ln\left(\frac{d}{r'}\right) \times 10^{-7} \text{ [H / m]} \tag{5.27}$$

In the above equations, d is the distance between the centers of conductors, and r is the radius of each of the conductors.

The capacitance of a single-phase two-wire line (Figure 5.12a) is:

$$C_{L-L} = C_{12} = \frac{\pi\varepsilon_0}{\ln \frac{d}{r}} \text{ [F / m]} \tag{5.28}$$

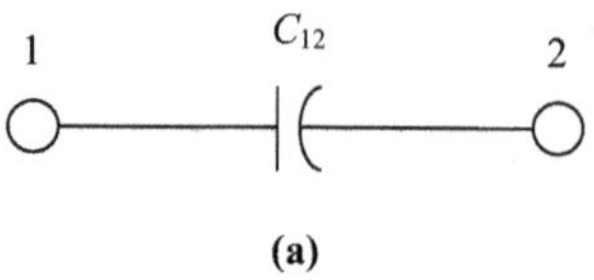

C_{12}

1 2

(a)

C C

1 N 2

(b)

FIGURE 5.12 (a) Line-to-line capacitance of a single-phase two-wire line. (b) Line-to-neutral capacitance of a single-phase two-wire line [18].

In the above equation, $\varepsilon_0 = 8.854 \times 10^{12}$ $[\text{F/m}]$. The above equation gives the line-to-line capacitance between the conductors. For the purpose of transmission line modeling, it is required to define a capacitance between each conductor and hypothetical neutral as shown in Figure 5.12b. Since the voltage to neutral is half of the line-to-line voltage, the capacitance to neutral is $C = 2C_{L-L}$. We thus have:

$$C = \frac{2\pi\varepsilon_0}{\ln \frac{d}{r}} \; [\text{F/m}] \tag{5.29}$$

In the case of three-phase systems, power flow analysis and balanced-fault analysis are usually performed on one phase only. From the system voltage and current results obtained for one phase, the voltages and currents in the other two phases are easily calculated since they have identical magnitudes and a progressive phase difference of 120°. For a three-phase transmission line with equilateral spacing, as shown in Figure 5.13, the inductance and capacitance are determined with respect to the hypothetical neutral conductor.

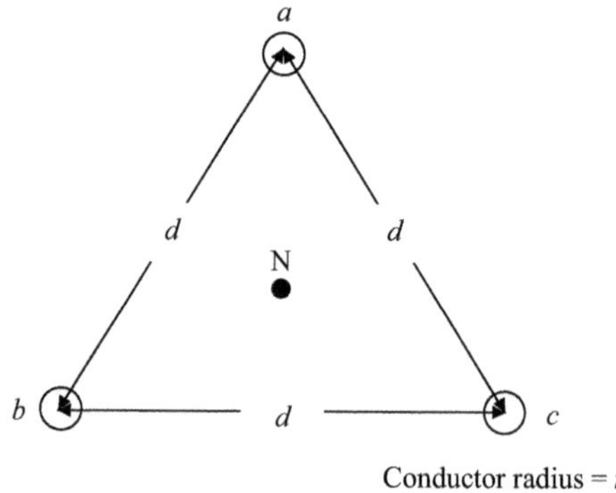

FIGURE 5.13 Overhead power transmission line with equilateral spacing between phase conductors [48].

The line-to-neutral inductance per phase per m length of a three-phase transmission line is:

$$L = 2\ \ln\left(\frac{d}{r'}\right)\times 10^{-7}\ [\mathrm{H/m}] \tag{5.30}$$

With three-phase conductors spaced equilaterally, the capacitance of each line to the hypothetical neutral is:

$$C = \frac{2\pi\varepsilon_0}{\ln\frac{d}{r}}\ [\mathrm{F/m}] \tag{5.31}$$

The three-phase conductors rarely have equilateral spacing. It can, however, be proved that for any other spacing of three-phase conductors, as shown in Figure 5.14, the average value of inductance or capacitance can be obtained by representation of the system by one having equivalent equilateral spacing. The equivalent spacing d_{EQ} between conductors is obtained using the following equation:

$$d_{\mathrm{EQ}} = \sqrt[3]{d_{12}\times d_{23}\times d_{31}} \tag{5.32}$$

The equivalent spacing d_{EQ} is used in place of d in equations (5.30) and (5.31) when conductors in three-phase systems are not equilaterally spaced.

Unsymmetrical conductor spacing results in different inductance in each phase. This leads to an unbalanced voltage drop in each phase, even for balanced three-phase load currents. The resultant voltage or current results in inducing undesirable voltages in nearby communication cables. To regain symmetry and have equal inductance in each phase of the transmission line, transposition of conductors is carried out as shown in Figure 5.15. Transposition involves interchanging phase conductor positions at regular intervals of one-third of the total transmission line length.

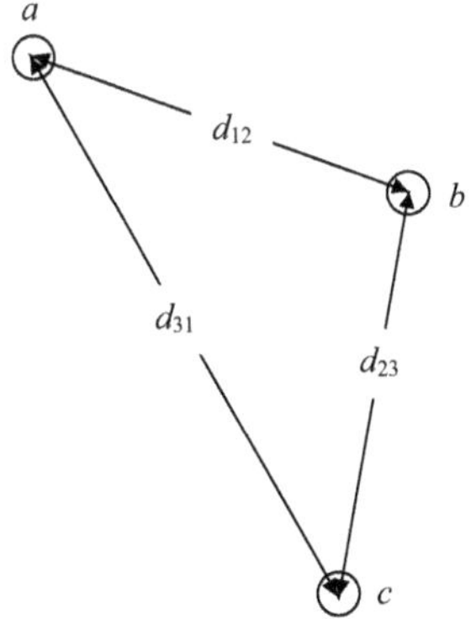

FIGURE 5.14 Transmission line conductors with unsymmetrical spacing [48].

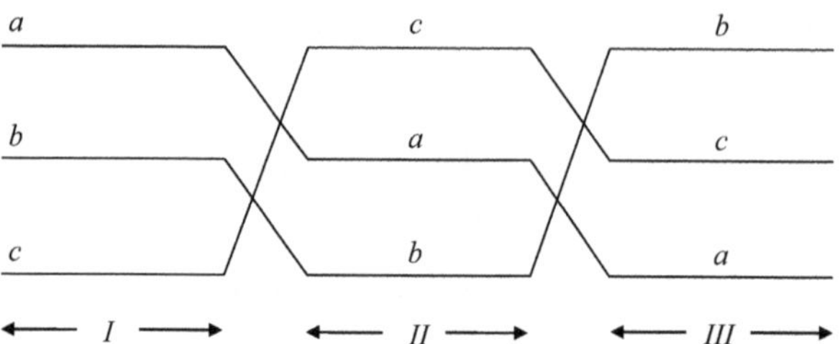

FIGURE 5.15 Transposed transmission line conductors [48].

Transposition is generally not carried out at regular intervals in practice, and transmission line conductors are transposed at locations where it is convenient to do so such as substations. The unbalance occurring in phase conductor inductance without transposition is small and is generally ignored. The transmission line conductors are always assumed to be transposed for the purpose of modeling [14,48].

Bundling of phase conductors or having more than one conductor per phase results in a reduction in the transmission line reactance. The student reader is requested to refer to standard power system references to learn about bundled conductors.

Example 5.3

The conductors of a 20 [km] long, 345 [kV] overhead transmission line are arranged horizontally with 1000 [mm] spacing between conductor centers. The effective diameter of the conductors is 40 [mm]. The resistance per kilometer of the conductors is 0.10 [Ω]. Determine the line-to-neutral inductance of the line and transmission line series impedance.

Solution

The equivalent equilateral spacing is given by:

$$d_{\mathrm{EQ}} = \sqrt[3]{d_{12} \times d_{23} \times d_{31}}$$

In this case, we have:

$$d_{\mathrm{EQ}} = \sqrt[3]{1000 \times 1000 \times 2000} = 1259.92\ [\mathrm{mm}]$$

$$r = 20\ [\mathrm{mm}]$$

The line-to-neutral inductance is:

$$L = 2\ \ln\left(\frac{d_{\mathrm{EQ}}}{r'}\right) \times 10^{-7}\ [\mathrm{H/m}]$$

$$= 2\ \ln\left(\frac{d_{\mathrm{EQ}}}{re^{-1/4}}\right) \times 10^{-7}\ [\mathrm{H/m}]$$

$$= 2\ \ln\left(\frac{1259.92}{20 \times e^{-1/4}}\right) \times 10^{-7}\ [\mathrm{H/m}]$$

$$= 0.7786 \times 10^{-6}\ [\mathrm{H/m}]$$

The total inductance of 20 [km] transmission line is:

$$L_{\text{total}} = 15.57\ [\mathrm{mH}]$$

The inductive reactance is:

$$X_L = 2\pi f L_{\text{total}} = 2\pi \times 60 \times 15.57 \times 10^{-3} = 5.87\ [\Omega]$$

Resistance of transmission line is:

$$R = 0.10\ [\Omega/\mathrm{km}]$$

Total resistance of transmission line is:

$$R_{total} = 20 \times 0.10 = 2\ [\Omega]$$

Impedance of the transmission line is:

$$Z = 2 + j5.87\ [\Omega]$$

5.12 TRANSMISSION LINE MODELS

In power system modeling and analysis, transmission lines are modeled using per-phase equivalent circuit representation with circuit parameters such as resistance, conductance, inductance, and capacitance specified on a per-phase basis [11,18,44,48]. The transmission line voltages in per-phase analysis are from line to neutral, and current is phase current. The three-phase system is therefore converted to a single-phase system for simplicity in analysis.

The model used to represent the transmission line depends upon the length of the transmission line. Transmission lines are broadly classified as short-length transmission lines for lengths up to 80 [km], medium-length transmission lines for lengths between 80 [km] and 240 [km], and long-length transmission lines for lengths above 240 [km]. In the following subsections, circuit parameters and voltage and current relations are derived for each of the three classifications of transmission lines mentioned above.

5.12.1 Short-Length Transmission Line Model

In the case of short transmission lines, capacitance and shunt conductance are generally neglected without much error. The short transmission line model as shown

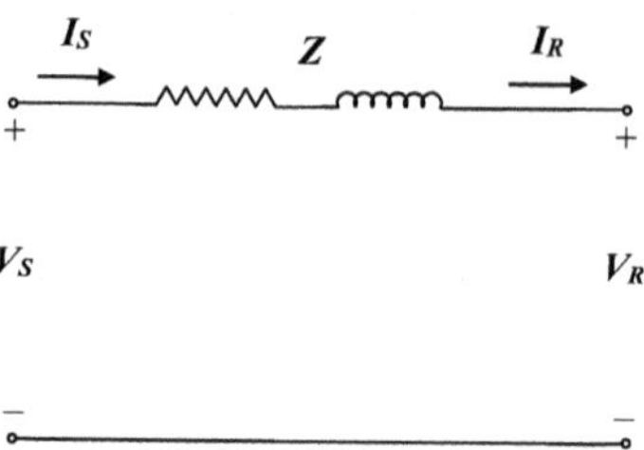

FIGURE 5.16 Short-length transmission line model.

in Figure 5.16 is obtained by multiplying series resistance and inductive reactance per-phase per-unit length with the length of the transmission line.

$$Z = (R_l + j\omega L_l)l = (R + j\omega L) = (R + jX) \tag{5.33}$$

In the above equation, R_l is the transmission line resistance per-phase per-unit length, L_l is the transmission line inductance per-phase per-unit length, and l is the length of the transmission line.

$\boldsymbol{V_S}$ and $\boldsymbol{I_S}$ in the transmission line model are phase voltage and current measured at the sending end of the transmission line. $\boldsymbol{V_R}$ and $\boldsymbol{I_R}$ are the phase voltage and current measured at the receiving end of the transmission line.

If there is a three-phase load $\boldsymbol{S_{R,3\phi}}$ connected at the receiving end of the transmission line, $\boldsymbol{I_R}$ can be calculated from the following equation:

$$\boldsymbol{I_R} = \frac{\boldsymbol{S}^*_{R,3\phi}}{3\boldsymbol{V}^*_R} \tag{5.34}$$

We therefore have:

$$\boldsymbol{V_S} = \boldsymbol{V_R} + \boldsymbol{Z}\boldsymbol{I_R} \tag{5.35}$$

$$\boldsymbol{I_S} = \boldsymbol{I_R} \tag{5.36}$$

It is possible to represent a short transmission line as a two-port or four-terminal network as shown in Figure 5.17.

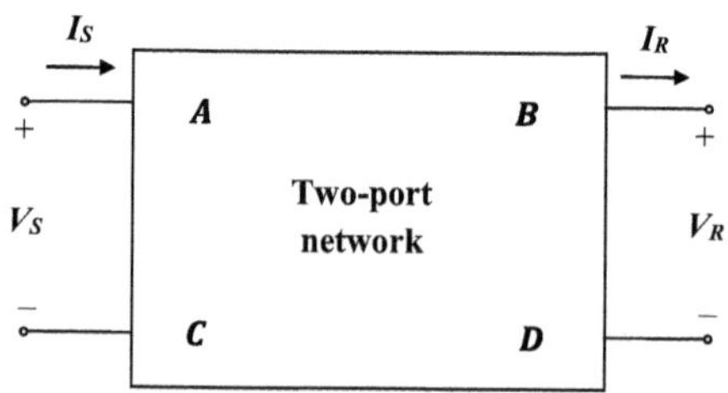

FIGURE 5.17 Transmission line representation as a two-port network.

We can write the voltage and current equations above for a short transmission line in terms of generalized constants or ***ABCD*** parameters of a two-port network as follows:

$$\boldsymbol{V_S} = \boldsymbol{A}\boldsymbol{V_R} + \boldsymbol{B}\boldsymbol{I_R} \tag{5.37}$$

$$\boldsymbol{I_S} = \boldsymbol{C}\boldsymbol{V_R} + \boldsymbol{D}\boldsymbol{I_R} \tag{5.38}$$

Using matrix notation, we have:

$$\begin{bmatrix} \boldsymbol{V_S} \\ \boldsymbol{I_S} \end{bmatrix} = \begin{bmatrix} \boldsymbol{A} & \boldsymbol{B} \\ \boldsymbol{C} & \boldsymbol{D} \end{bmatrix} \begin{bmatrix} \boldsymbol{V_R} \\ \boldsymbol{I_R} \end{bmatrix} \tag{5.39}$$

From equations (5.35) and (5.36) above for the short transmission line model, we find:

$$\boldsymbol{A} = 1 \quad \boldsymbol{B} = \boldsymbol{Z} \quad \boldsymbol{C} = 0 \quad \boldsymbol{D} = 1 \tag{5.40}$$

The sending-end three-phase power $\boldsymbol{S}_{S,3\phi}$ is:

$$\boldsymbol{S}_{S,3\phi} = 3\boldsymbol{V}_S\boldsymbol{I}_S^* \tag{5.41}$$

The transmission line loss $\boldsymbol{S}_{\mathbf{Loss},3\phi}$ is:

$$\boldsymbol{S}_{\mathbf{Loss},3\phi} = \boldsymbol{S}_{S,3\phi} - \boldsymbol{S}_{R,3\phi} \tag{5.42}$$

The efficiency of the transmission line is:

$$\eta = \frac{P_{R,3\phi}}{P_{S,3\phi}} \tag{5.43}$$

In the above equation, $P_{R,3\phi}$ is the real power at the receiving end of the transmission line, whereas $P_{S,3\phi}$ is the real power at the sending end of the transmission line.

5.12.2 Medium-Length Transmission Line Model

As transmission lines become longer, calculation of voltages, currents, and power flows using equations developed for the short-length transmission line model provides inaccurate results. The effect of current leaking to ground through shunt capacitance is therefore considered for improved approximation. Transmission lines having lengths between 80 and 240 [km] are called medium-length transmission lines. Medium-length transmission lines have half of the total shunt capacitive reactance lumped at each end of the line forming π network as shown in Figure 5.18. The total shunt admittance $\boldsymbol{Y}$ of the medium-length transmission line is:

$$\boldsymbol{Y} = \left(G_l + j\omega C_l\right)l = \left(G + j\omega C\right) \tag{5.44}$$

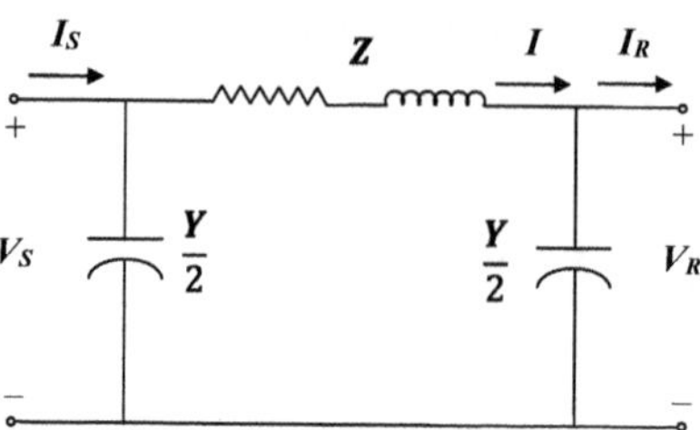

FIGURE 5.18 π model for a medium-length transmission line.

In the above equation, G_l is the transmission line shunt conductance per-phase per-unit length, C_l is the transmission line capacitance per-phase per-unit length, and l is the length of the transmission line. The shunt conductance per-unit length G_l is negligible, and it is generally neglected.

The total series impedance of the medium-length transmission line is determined using equation (5.33) as described earlier.

$$\boldsymbol{Z} = (R_l + j\omega L_l)l = (R + j\omega L) = (R + jX) \tag{5.33}$$

From Figure 5.18, we have:

$$\boldsymbol{I} = \boldsymbol{I}_R + \frac{\boldsymbol{Y}}{2}\boldsymbol{V}_R \tag{5.45}$$

$$\boldsymbol{V}_S = \boldsymbol{V}_R + \boldsymbol{ZI} \tag{5.46}$$

After substituting $\boldsymbol{I}$ from equation (5.45) in equation (5.46), we obtain:

$$\boldsymbol{V}_S = \left(1 + \frac{\boldsymbol{ZY}}{2}\right)\boldsymbol{V}_R + \boldsymbol{ZI}_R \tag{5.47}$$

The current at the sending end $\boldsymbol{I}_S$ is:

$$\boldsymbol{I}_S = \boldsymbol{I} + \frac{\boldsymbol{Y}}{2}\boldsymbol{V}_S \tag{5.48}$$

After substituting $\boldsymbol{I}$ from equation (5.45) and $\boldsymbol{V}_S$ from equation (5.47) in equation (5.48), we obtain:

$$\boldsymbol{I}_S = \boldsymbol{Y}\left(1 + \frac{\boldsymbol{ZY}}{4}\right)\boldsymbol{V}_R + \left(1 + \frac{\boldsymbol{ZY}}{2}\right)\boldsymbol{I}_R \tag{5.49}$$

Equations (5.47) and (5.49) above express sending-end voltage and current in terms of receiving-end voltage and current. Similar to the short-length transmission line, we can obtain $\boldsymbol{ABCD}$ parameters for the medium-length transmission line. Using equations (5.37), (5.38), (5.47), and (5.49), we obtain:

$$\boldsymbol{A}=\left(1+\frac{\boldsymbol{ZY}}{2}\right) \quad \boldsymbol{B}=\boldsymbol{Z} \tag{5.50}$$

$$\boldsymbol{C}=\boldsymbol{Y}\left(1+\frac{\boldsymbol{ZY}}{4}\right) \quad \boldsymbol{D}=\left(1+\frac{\boldsymbol{ZY}}{2}\right) \tag{5.51}$$

5.12.3 Long-Length Transmission Line Model

Sufficiently accurate models were obtained earlier for short-length and medium-length transmission lines by assuming the line parameters to be lumped. For transmission lines longer than 240 [km], more accurate analysis requires line parameters to be distributed uniformly throughout the length of the transmission line and not to lump the line parameters [11,18,44,48].

For the long-length transmission line, we first derive equations to determine voltage and current at any location on the line. An equivalent π model is subsequently developed for the long-length transmission line using these voltage and current equations obtained. Figure 5.19 shows one phase of a long-length transmission line of length l with distributed parameters z in this figure is the series impedance per-unit length, and it equals $(R_l + j\omega L_l)$, whereas $\boldsymbol{y}$ is the shunt admittance of the transmission line per-unit length, and it equals $(G_l + j\omega C_l)$.

A transmission line element of length Δx located at a distance x from the receiving end of the transmission line is considered for the derivation of voltage and current equations. The voltages and currents at the two ends of this section are shown as a function of distance. We can write the following voltage equation for the elemental section:

$$\boldsymbol{V}(\boldsymbol{x}+\Delta\boldsymbol{x})-\boldsymbol{V}(\boldsymbol{x})=z\Delta x\boldsymbol{I}(x)$$

$$\frac{\boldsymbol{V}(\boldsymbol{x}+\Delta\boldsymbol{x})-\boldsymbol{V}(\boldsymbol{x})}{\Delta\boldsymbol{x}}=z\boldsymbol{I}(x) \tag{5.52}$$

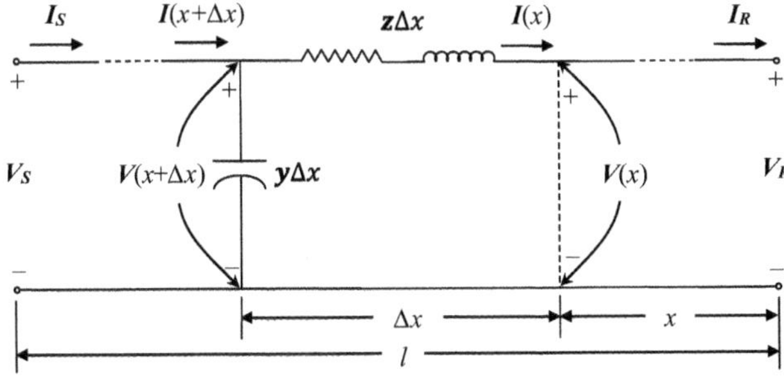

FIGURE 5.19 Long-length transmission line with distributed parameters [XVIII].

As Δx tends to 0, we obtain:

$$\frac{d\,\boldsymbol{V}(\boldsymbol{x})}{d\boldsymbol{x}} = z\boldsymbol{I}(x) \tag{5.53}$$

We can also write the following current equation for the elemental section:

$$\boldsymbol{I}(\boldsymbol{x}+\Delta\boldsymbol{x}) - \boldsymbol{I}(\boldsymbol{x}) = \boldsymbol{y}\Delta x\boldsymbol{V}(x+\Delta x)$$

$$\frac{\boldsymbol{I}(\boldsymbol{x}+\Delta\boldsymbol{x}) - \boldsymbol{I}(\boldsymbol{x})}{\Delta\boldsymbol{x}} = \boldsymbol{y}\boldsymbol{V}(x+\Delta x) \tag{5.54}$$

As Δx tends to 0, we similarly obtain:

$$\frac{d\,\boldsymbol{I}(\boldsymbol{x})}{d\boldsymbol{x}} = \boldsymbol{y}\boldsymbol{V}(x) \tag{5.55}$$

Differentiating equation (5.53) and substituting into equation (5.55), we obtain:

$$\frac{d^2\boldsymbol{V}(\boldsymbol{x})}{d\boldsymbol{x}^2} = zy\boldsymbol{V}(x) \tag{5.56}$$

We replace zy in the above equation with another parameter $\mathcal{P}^2$ so that $\mathcal{P}^2 = zy$. $\mathcal{P} = \sqrt{zy}$ is called propagation constant per-unit length. We can write:

$$\mathcal{P} = \sqrt{zy} = \alpha + j\beta = \sqrt{(R_l + j\omega L_l)(G_l + j\omega C_l)} \tag{5.57}$$

The real part α in the above complex expression is called the attenuation constant, and the imaginary part β is called the phase constant.

We therefore have:

$$\frac{d^2\boldsymbol{V}(\boldsymbol{x})}{d\boldsymbol{x}^2} - \mathcal{P}^2\boldsymbol{V}(\boldsymbol{x}) = 0 \tag{5.58}$$

The above second-order differential equation has the following solution:

$$\boldsymbol{V}(\boldsymbol{x}) = \boldsymbol{C}_1\boldsymbol{e}^{\mathcal{P}x} + \boldsymbol{C}_2\boldsymbol{e}^{-\mathcal{P}x} \tag{5.59}$$

From equation (5.53), we obtain:

$$\boldsymbol{I}(\boldsymbol{x}) = \frac{1}{z}\frac{d\boldsymbol{V}(\boldsymbol{x})}{d\boldsymbol{x}} = \frac{\mathcal{P}}{z}\left(\boldsymbol{C}_1\boldsymbol{e}^{\mathcal{P}x} - \boldsymbol{C}_2\boldsymbol{e}^{-\mathcal{P}x}\right) = \sqrt{\frac{y}{z}}\left(\boldsymbol{C}_1\boldsymbol{e}^{\mathcal{P}x} - \boldsymbol{C}_2\boldsymbol{e}^{-\mathcal{P}x}\right) \tag{5.60}$$

We replace $\sqrt{\frac{z}{y}}$ in the above equation with another parameter $\boldsymbol{Z}_c$ so that $\boldsymbol{Z}_c = \sqrt{\frac{z}{y}}$. $\boldsymbol{Z}_c$ is called the characteristic impedance of the transmission line per-unit length.

We therefore obtain:

$$\boldsymbol{I}(\boldsymbol{x}) = \frac{1}{\boldsymbol{Z}_c}\left(\boldsymbol{C}_1 e^{\mathcal{P}x} - \boldsymbol{C}_2 e^{-\mathcal{P}x}\right) \tag{5.61}$$

We next determine constants $\boldsymbol{C}_1$ and $\boldsymbol{C}_2$. For the transmission line shown in Figure 5.19, we have for $x = 0$: $\boldsymbol{V}(\boldsymbol{x}) = \boldsymbol{V}_R$ and $\boldsymbol{I}(\boldsymbol{x}) = \boldsymbol{I}_R$. Substituting these values in equations (5.59) and (5.61) provides the following values of $\boldsymbol{C}_1$ and $\boldsymbol{C}_2$:

$$\boldsymbol{C}_1 = \frac{\boldsymbol{V}_R + \boldsymbol{Z}_c\boldsymbol{I}_R}{2} \tag{5.62}$$

$$\boldsymbol{C}_2 = \frac{\boldsymbol{V}_R - \boldsymbol{Z}_c\boldsymbol{I}_R}{2} \tag{5.63}$$

From equations (5.59–5.63), we obtain the following equations for voltage and current at any location on the transmission line:

$$\boldsymbol{V}(\boldsymbol{x}) = \frac{\boldsymbol{V}_R + \boldsymbol{Z}_c\boldsymbol{I}_R}{2} e^{\mathcal{P}x} + \frac{\boldsymbol{V}_R - \boldsymbol{Z}_c\boldsymbol{I}_R}{2} e^{-\mathcal{P}x} \tag{5.64}$$

$$\boldsymbol{I}(\boldsymbol{x}) = \frac{1}{\boldsymbol{Z}_c}\left(\frac{\boldsymbol{V}_R + \boldsymbol{Z}_c\boldsymbol{I}_R}{2} e^{\mathcal{P}x} - \frac{\boldsymbol{V}_R - \boldsymbol{Z}_c\boldsymbol{I}_R}{2} e^{-\mathcal{P}x}\right) = \frac{\frac{\boldsymbol{V}_R}{\boldsymbol{Z}_c} + \boldsymbol{I}_R}{2} e^{\mathcal{P}x} - \frac{\frac{\boldsymbol{V}_R}{\boldsymbol{Z}_c} - \boldsymbol{I}_R}{2} e^{-\mathcal{P}x} \tag{5.65}$$

We can rearrange the above equations as follows:

$$\boldsymbol{V}(\boldsymbol{x}) = \frac{e^{\mathcal{P}x} + e^{-\mathcal{P}x}}{2} \boldsymbol{V}_R + \boldsymbol{Z}_c \frac{e^{\mathcal{P}x} - e^{-\mathcal{P}x}}{2} \boldsymbol{I}_R \tag{5.66}$$

$$\boldsymbol{I}(\boldsymbol{x}) = \frac{1}{\boldsymbol{Z}_c} \frac{e^{\mathcal{P}x} - e^{-\mathcal{P}x}}{2} \boldsymbol{V}_R + \frac{e^{\mathcal{P}x} + e^{-\mathcal{P}x}}{2} \boldsymbol{I}_R \tag{5.67}$$

In terms of hyperbolic functions, the above equations become:

$$\boldsymbol{V}(\boldsymbol{x}) = \cosh \mathcal{P}x \boldsymbol{V}_R + \boldsymbol{Z}_c \sinh \mathcal{P}x \boldsymbol{I}_R \tag{5.68}$$

$$\boldsymbol{I}(\boldsymbol{x}) = \frac{1}{\boldsymbol{Z}_c} \sinh \mathcal{P}x \boldsymbol{V}_R + \cosh \mathcal{P}x \boldsymbol{I}_R \tag{5.69}$$

From the above equations, we can determine voltage and current at any location on the long transmission line.

We next determine the relationship between sending-end quantities and receiving-end quantities. From Figure 5.19, we have for $x = l$: $\boldsymbol{V}(\boldsymbol{x}) = \boldsymbol{V}_S$ and $\boldsymbol{I}(\boldsymbol{x}) = \boldsymbol{I}_S$. Substituting these values in equations (5.68) and (5.69), we obtain:

$$V_S = \cosh \mathcal{P}l V_R + Z_c \sinh \mathcal{P}l I_R \tag{5.70}$$

$$I_S = \frac{1}{Z_c} \sinh \mathcal{P}l V_R + \cosh \mathcal{P}l I_R \tag{5.71}$$

Similar to short-length and medium-length transmission lines, we can obtain ***ABCD*** parameters for the long-length transmission line. We rewrite the above equations in matrix form as follows:

$$\begin{bmatrix} V_S \\ I_S \end{bmatrix} = \begin{bmatrix} A & B \\ C & D \end{bmatrix} \begin{bmatrix} V_R \\ I_R \end{bmatrix} \tag{5.39}$$

We therefore have from equations (5.39), (5.70), and (5.71):

$$A = \cosh \mathcal{P}l \quad B = Z_c \sinh \mathcal{P}l \tag{5.72}$$

$$C = \frac{1}{Z_c} \sinh \mathcal{P}l \quad D = \cosh \mathcal{P}l \tag{5.73}$$

We next determine parameters of the equivalent π model for the long-length transmission line. Similar to equations (5.47) and (5.49) obtained for the medium-length transmission line π model, we can rewrite equations (5.70) and (5.71) for the long-length transmission line as follows:

$$V_S = \left(1 + \frac{Z'Y'}{2}\right) V_R + Z' I_R \tag{5.74}$$

$$I_S = Y'\left(1 + \frac{Z'Y'}{4}\right) V_R + \left(1 + \frac{Z'Y'}{2}\right) I_R \tag{5.75}$$

where

$$Z' = Z_c \sinh \mathcal{P}l \tag{5.76}$$

$$\frac{Y'}{2} = \frac{1}{Z_c} \frac{\cosh \mathcal{P}l - 1}{\sinh \mathcal{P}l} = \frac{1}{Z_c} \tanh \frac{\mathcal{P}l}{2} \tag{5.77}$$

Equations (5.74) and (5.75) have different forms; however, they are equivalent to equations (5.70) and (5.71), and they provide exactly similar values for sending-end voltage and current.

The equivalent π model of the long-length transmission line is shown in Figure 5.20.

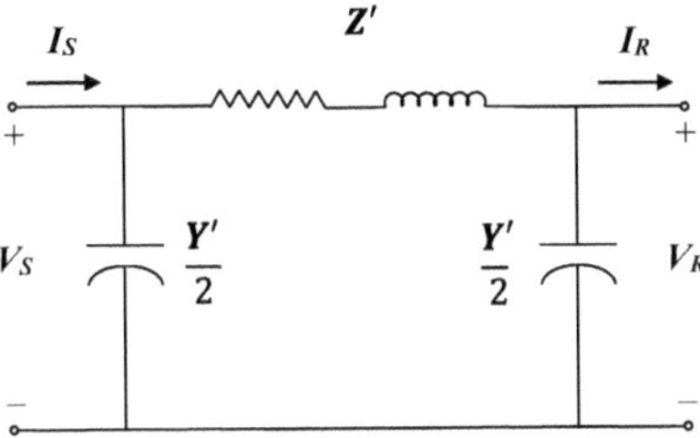

FIGURE 5.20 Long-length transmission line equivalent π model [18].

Example 5.4

A 345 [kV] three-phase transmission line is 20 [km] long. The resistance per phase is 0.14 [Ω/km], and the inductance per phase is 1.0075 [mH/km]. The shunt capacitance is negligible.

Using the short transmission line model, determine the voltage and power at the sending end and transmission efficiency when the line is supplying a three-phase load of 100 [MVA] at 0.95 power factor lagging at 345 [kV].

Solution

The series impedance per phase is:

$$\boldsymbol{Z} = \left(R_l + j\omega L_l\right)l = \left(0.14 + j2\pi\times 60\times 1.0075\times 10^{-3}\right)\times 20 = 2.8 + j7.6\ [\Omega]$$

The receiving-end voltage per phase is:

$$\boldsymbol{V_R} = \frac{345\angle 0^\circ}{\sqrt{3}} = 199.2\angle 0^\circ\ [\text{kV}]$$

Apparent power at the receiving end is:

$$\boldsymbol{S_{R,3\phi}} = 100\angle\cos^{-1}0.95 = 100\angle 18.2^\circ = 95 + j31.2\ [\text{MVA}]$$

The current per phase at the receiving end is:

$$\boldsymbol{I_R} = \frac{\boldsymbol{S^*_{R,3\phi}}}{3\boldsymbol{V^*_R}} = \frac{100\angle -18.2^\circ\times 10^3}{3\times 199.2\angle 0^\circ} = 167.3\angle -18.2^\circ\ [\text{A}]$$

The sending-end phase voltage is:

$$\boldsymbol{V_S} = \boldsymbol{V_R} + \boldsymbol{Z}\boldsymbol{I_R} = 199.2\angle 0^\circ + \left(2.8 + j7.6\right)\left(167.3\angle -18.2^\circ\right)\left(10^{-3}\right)$$

$$= 200.04\angle 0.3^\circ\ [\text{kV}]$$

Sending-end line-to-line voltage magnitude is:

$$\left|\boldsymbol{V_{S,LL}}\right| = \sqrt{3}\left|\boldsymbol{V_S}\right| = 346.5\ [\text{kV}]$$

The sending-end power is:

$$S_{S,3\phi} = 3V_S I_S^* = 3\times 200.04\angle 0.3^\circ \times 167.3\angle 18.2^\circ \times 10^{-3}$$
$$= 95.24\ [\text{MW}] + j31.86\ [\text{MVAR}]$$
$$= 100.42\angle 18.5^\circ\ [\text{MVA}]$$

Transmission line efficiency is:

$$\eta = \frac{P_{R,3\phi}}{P_{S,3\phi}} = \frac{95}{95.24}\times 100 = 99.75\ [\%]$$

Part IV: Power System Loads

5.13 POWER SYSTEM LOAD MODELING

The main load types for utilization consist of industrial, commercial, and residential loads. Depending on the requirements of their activities, consumers consume small or large amounts of electrical power. Since consumer demands are unpredictable, a power system's load varies over time. The generation capacity of power plants should be capable of supplying the maximum possible demand requirement of the consumers.

Different power system loads mainly include static loads and dynamic loads. Static loads include equipment such as heating equipment, lighting equipment, and electronic equipment [34]. They are impedance-type loads, and their impedance is given by $\boldsymbol{Z} = R + jX$. Dynamic loads encountered in power system networks include motor devices such as synchronous motors and induction motors.

The single-phase equivalent circuit of a synchronous motor after ignoring the armature resistance is shown in Figure 5.21a, which is almost similar to the single-phase equivalent circuit of a synchronous generator described earlier in this chapter. The single-phase equivalent circuit of an induction motor is similar to that of a two-winding transformer. Figure 5.21b shows the equivalent circuit of an induction motor referred to as the stator. $\boldsymbol{V}_s$ is the stator phase rms voltage, and $\boldsymbol{I}_s$ is the current supplied to the stator winding. R_s and X_s in this figure denote stator resistance and reactance, respectively. R_r and X_r are rotor resistance and reactance. The exciting or magnetizing circuit is composed of R_O and X_O. a is the effective stator/rotor turns ratio ($a = N_s / N_r$). The variable resistance represents the load on the motor. s is the slip of the induction motor rotor. The magnetizing impedance $\boldsymbol{Z_O}$ consisting of R_O and X_O is generally neglected.

The reader is requested to refer to any standard electric machinery text to learn more about equivalent circuits of synchronous motor and induction motor loads.

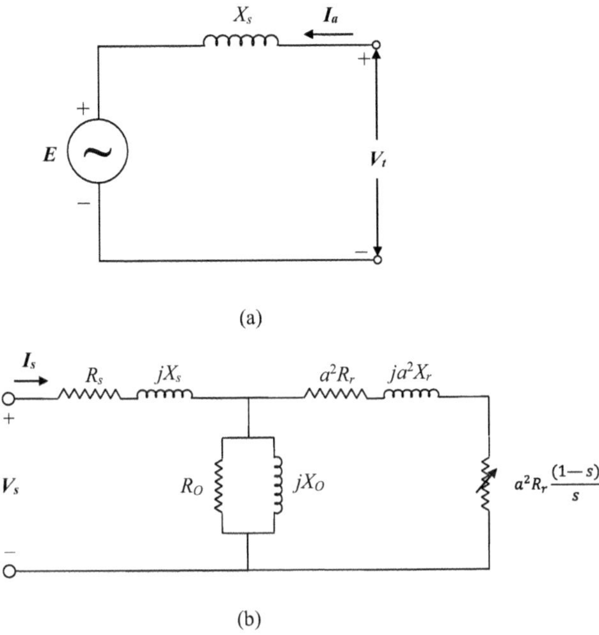

FIGURE 5.21 Single-phase equivalent circuits: (a) synchronous motor and (b) induction motor.

Example 5.5

A static load has a balanced Δ-connected impedance of $5\angle 45°$ [Ω per phase]. The load is connected through three balanced line impedances of $1 + j1$ [Ω] to a three-phase source that has a line-to-line voltage of 1000 [V]. Determine the magnitude of the line current delivered to the load.

Solution

$$|\mathbf{V}_{\mathrm{LL}}| = 1000\ [\mathrm{V}]$$

$$|\mathbf{V}_{\mathrm{LN}}| = \frac{1000}{\sqrt{3}} = 577.36\ [\mathrm{V}]$$

$$\mathbf{V}_{\mathrm{LN}} = 577.36\angle 0°$$

$$\mathbf{Z}_{\mathrm{TL}} = 1 + j1\ [\Omega]$$

$$\mathbf{Z}_{\mathrm{Load},\,\Delta} = 5\angle 45°\ [\Omega]$$

$$\mathbf{Z}_{\mathrm{Load},\,Y} = \frac{\mathbf{Z}_{\mathrm{Load,\,Z}}}{3} = \frac{5\angle 45°}{3} = 1.67\angle 45°\ [\Omega] = 1.18 + j1.18\ [\Omega]$$

$$\begin{aligned} Z_{EQ} &= \text{Per-phase equivalent impedance} \\ &= \text{Line impedance} + \text{equivalent wye impedance of the load} \end{aligned}$$

$$Z_{EQ} = Z_{TL} + Z_{Load,\,Y} = (1 + j1) + (1.18 + j1.18) = (2.18 + j2.18)\ [\Omega]$$

$$Z_{EQ} = 3.08\angle 45°\ [\Omega]$$

$$I_L = \frac{V_{LN}}{Z_{EQ}} = \frac{577.36\angle 0°}{3.08\angle 45°} = 187.46\angle -45°\ [\text{A}]$$

5.14 SUMMARY

The development of simplified equivalent circuits for the power system components such as synchronous generators, transformers, transmission lines, and system loads is described in the present chapter, and it is of great value in power system analysis.

During proper operation of the power system, voltage sources and loads are balanced in the three phases, and per-phase analysis can be used to determine system voltages and currents. The circuit models developed for different power system components are single-phase due to this reason. The system voltage and current results obtained for one phase of the power system can be easily used to determine voltages and currents in the other two phases.

REVIEW QUESTIONS

5.1 How is the electrical frequency of a synchronous generator related to the rate of rotation of the generator rotor?

5.2 Describe what the term armature reaction means and how it is modeled in the case of a synchronous generator.

5.3 Derive the equivalent circuit model of a synchronous generator.

5.4 Describe how a simplified circuit model of a synchronous generator is obtained from an equivalent circuit model.

5.5 Describe the procedure for synchronization of a synchronous generator to the power grid.

5.6 Describe the effects of variation in synchronous machine excitation on the supply of reactive power.

5.7 Describe the different types of losses which occur in a transformer.

5.8 Describe how the equivalent circuit of a real non-ideal transformer is determined.

5.9 Describe the different kinds of three-phase transformers and how they are analyzed.

5.10 Describe briefly the term skin effect and how it affects transmission line resistance.

5.11 What is the shunt conductance of a transmission line?

5.12 What is the equation for the per-phase inductance of a three-phase transmission line?

5.13 What is the equation for the per-phase capacitance of a three-phase transmission line?

5.14 Derive the matrix equation for the short-length transmission line relating receiving-end voltage and current with sending-end voltage and current.

5.15 Derive the matrix equation for the medium-length transmission line relating receiving-end voltage and current with sending-end voltage and current.

5.16 Derive the matrix equation for the long-length transmission line relating receiving-end voltage and current with sending-end voltage and current.

5.17 Describe how induction motor loads are modeled in power system analysis.

5.18 Describe how synchronous motor loads are modeled in power system analysis.

6 Power System Representation

6.1 INTRODUCTION

The different power system components were described in the previous chapter. These elements can be easily combined to create various designs of power systems. This chapter describes the representation of power systems by means of one-line diagrams and various symbols used in their construction. It also describes the application of the per-unit system to power systems, which enormously simplifies the solving of power system problems containing transformers. Real three-phase power systems containing many transformers are then easily analyzed using their per-phase per-unit equivalent circuits.

6.2 POWER SYSTEM ONE-LINE DIAGRAMS

A three-phase power system is always balanced with balanced voltage sources in each phase having equal voltage magnitudes and 120° phase difference connected through three similar line conductors to equal load impedances in the three phases. Since the three phases of the power system are exactly similar after ignoring the progressive phase difference, power systems are generally drawn in simplified form using standard symbols for various power system components and a three-phase transmission line represented by a single line. These one-line diagrams include all main components of the power system such as generators, transformers, transmission lines, and loads and describe how different power system components are arranged and interconnected in the real power system [6,13,14]. They provide a way of supplying much information for power system analysis work. A simple power system is shown in Figure 6.1 together with the corresponding one-line diagram.

Some standard symbols used in one-line diagrams are shown in Figure 6.2. Another one-line diagram of a more complex power system is shown in Figure 6.3. The power system has a synchronous generator supplying power, four buses, two transformers, a transmission line, a synchronous motor load, and a three-phase impedance load along with circuit breakers protecting all devices. Also shown in the power system one-line diagram are the type of equipment connections such as delta and wye connections, and the locations where power system components have connections made to the ground.

The one-line diagram should indicate where a power system is connected to ground, since these connections affect the current that will flow in power systems during ground faults. The ground connections are sometimes solid or without any intentional impedance in between, and sometimes through resistors or inductors. These resistors or inductors help in reducing the ground fault current in case of a

DOI: 10.1201/9781003432340-6

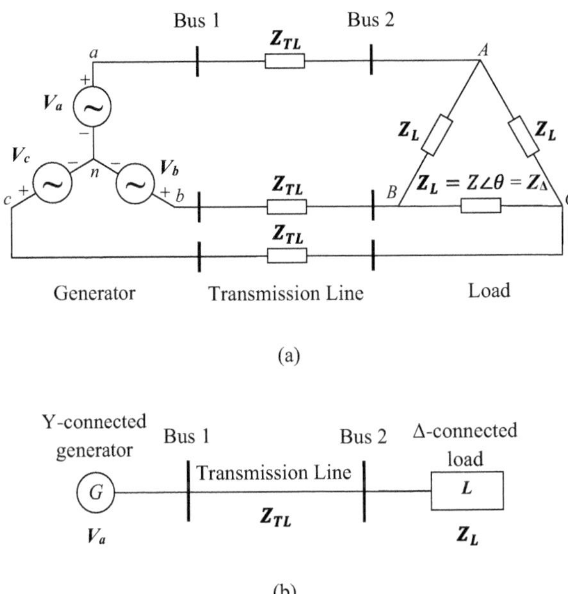

FIGURE 6.1 (a) Simple power system with a *Y*-connected generator, transmission line, and a Δ-connected load, and (b) corresponding one-line diagram.

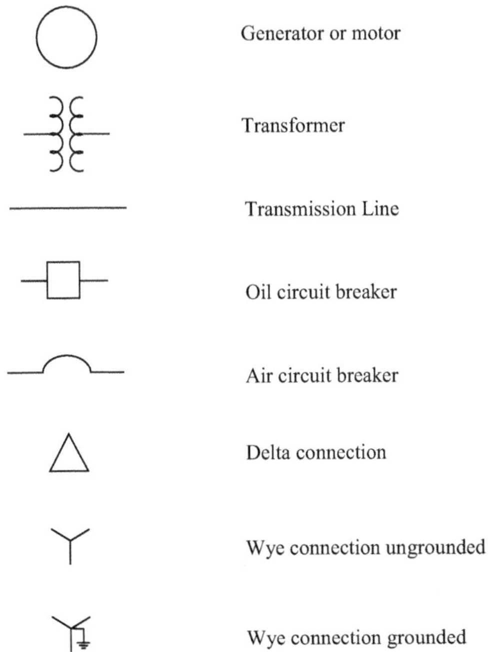

FIGURE 6.2 Symbols used in power system one-line diagrams.

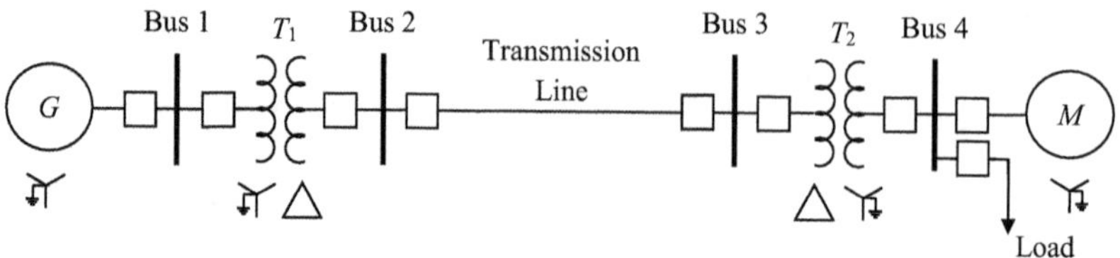

FIGURE 6.3 A one-line diagram of a power system.

ground fault such as single line-to-ground fault; however, they do not affect the working of the balanced three-phase power system, since the current flowing in neutral of a balanced three-phase system is zero. The relevant data pertaining to various components such as equipment power ratings, voltages, resistances, and reactances is also many times provided on the one-line diagram [6].

Finally, the purpose of the one-line diagram determines the amount of information to be shown. For example, if the diagram is for power flow analysis described in a different chapter later, the location of circuit breakers and relays is not necessary. However, in a power system protection study, knowledge of such devices is required. It must also be remembered that one-line diagrams contain only known facts about the power system, and any implications or guessing of information not supplied can have disastrous consequences [35].

6.3 PER-UNIT SYSTEM

Power system problems containing many transformers and thus several different voltage levels are solved after a difficult transformation process that involves bringing all system impedances to a single voltage level. Another approach to solving power system problems is the per-unit method in which various power system quantities such as voltage, current, power, and impedance are expressed as decimal fractions of corresponding base quantities. In the per-unit method of calculations, different voltage levels in a power system due to the presence of transformers disappear, and a power system network consisting of various power system components operating at different voltage levels is reduced to a system of impedances.

The per-unit value of a quantity is calculated using the following equation:

$$\text{Quantity in per-unit} = \frac{\text{Actual value}}{\text{Base value of quantity}} \tag{6.1}$$

The actual value of the quantity is its value in actual units, such as V, A, W/VAR/VA, and Ω. The base value of the quantity is a real number, and it has the same units as the actual value. The per-unit value of a quantity is therefore dimensionless, and it has the same phase angle as the actual quantity.

Two arbitrary base values out of four base values of base apparent power, base voltage, base current, and base impedance are selected at a particular location in a power system to convert actual values to per-unit system values. Generally, base apparent power and base voltage values are selected. The other base values of current and impedance are related to the selected base values by usual electrical laws.

In case of single-phase systems, we assume S_{base}, P_{base}, Q_{base}, and V_{base} to be base apparent power, base real power, base reactive power, and base voltage, respectively. The following relationships are then used for determining other base values in single-phase systems:

$$P_{base},\ Q_{base} \text{ or } S_{base} = \text{base voltage} \times \text{base current} = V_{base} \times I_{base}\,[\text{W/VAR/VA}] \tag{6.2}$$

$$I_{base} = \text{base current} = \frac{S_{base}}{V_{base}}\ [\text{A}] \tag{6.3}$$

$$Z_{base} = \text{base impedance} = \frac{V_{base}}{I_{base}} = \frac{(V_{base})^2}{S_{base}}\ [\Omega] \tag{6.4}$$

Base apparent power and base voltage are initially chosen at a particular location in the power system. The base apparent power is the same for the entire power system containing one or more transformers since the apparent power entering a transformer equals the apparent power coming out of it. The ratio of voltage bases on the two sides of a transformer is selected to be the same as its turns ratio. The following equations are then used for determining per-unit quantities:

$$\text{Per-unit voltage} = \frac{\text{actual voltage}}{\text{base voltage}}\ [\text{per-unit or pu}] \tag{6.5}$$

$$\text{Per-unit current} = \frac{\text{actual current}}{\text{base current}}\ [\text{per-unit or pu}] \tag{6.6}$$

$$\text{Per-unit impedance} = \frac{\text{actual impedance}}{\text{base impedance}}\ [\text{per-unit or pu}] \tag{6.7}$$

The above requirements for base apparent power and base voltage ensure that per-unit impedance remains unchanged when referred from one side of the transformer to the other.

The above description for single-phase system per-unit quantities can easily be extended to the three-phase systems. The single-phase base equations described above are applicable to three-phase systems on per-phase basis as follows:

$$I_{base} = \frac{S_{base,\,1\phi}}{V_{base,\,LN}}\ [\text{A}] \tag{6.8}$$

$$Z_{base} = \frac{V_{base,\,LN}}{I_{base}} = \frac{(V_{base,\,LN})^2}{S_{base,\,1\phi}}\ [\Omega] \tag{6.9}$$

In the above equations, $S_{base,1\phi}$ is the per-phase base apparent power and $V_{base,LN}$ is the base line-to-neutral voltage for the three-phase system.

The base values for three-phase systems can also be selected as three-phase basis. Generally, $S_{base,3\phi}$ and $V_{base,LL}$ are selected where $S_{base,3\phi}$ is three-phase base apparent power and $V_{base,LL}$ is base line-to-line voltage. The following relationships are then used for determining other base values in three-phase systems:

$$S_{base,1\phi} = \frac{S_{base,3\phi}}{3} \text{ [VA]} \quad (6.10)$$

$$V_{base,LN} = \frac{V_{base,LL}}{\sqrt{3}} \text{ [V]} \quad (6.11)$$

$$I_{base} = \frac{S_{base,3\phi}}{\sqrt{3}V_{base,LL}} \text{ [A]} \quad (6.12)$$

$$Z_{base} = \frac{V_{base,LL}}{\sqrt{3}I_{base}} = \frac{\left(V_{base,LL}\right)^2}{S_{base,3\phi}} \text{ [}\Omega\text{]} \quad (6.13)$$

When there is only one power system component such as a generator to be considered for per-unit analysis, its nameplate ratings can be used as base values. In the per-unit analysis of power system having many components, the base values selected will be different from nameplate ratings of a particular component. It will then be necessary to convert per-unit values of the component from its nameplate rating to selected power system base values. The relationships between old and new per-unit values are as follows:

$$\left[P,\ Q,\ S\right]_{pu\ on\ new\ base} = \left[P,\ Q,\ S\right]_{pu\ on\ old\ base} \times \frac{S_{old\ base}}{S_{new\ base}} \quad (6.14)$$

$$V_{pu\ on\ new\ base} = V_{pu\ on\ old\ base} \times \frac{V_{old\ base}}{V_{new\ base}} \quad (6.15)$$

$$\left[R,\ X,\ Z\right]_{pu\ on\ new\ base} = \left[R,\ X,\ Z\right]_{pu\ on\ old\ base} \times \frac{V_{old\ base}^{2}}{V_{new\ base}^{2}} \times \frac{S_{new\ base}}{S_{old\ base}} \quad (6.16)$$

In the above equations, subscripts old base and new base refer to old and new base values, respectively.

Example 6.1

A simple single-phase power system has a base impedance value of 100 [Ω] and a base voltage value of 600 [V]. Determine the base current and base apparent power values of the power system.

Solution

We have

$$I_{\text{base}} = \text{Base current} = \frac{\text{base voltage}}{\text{base impedance}} = \frac{600}{100} = 6\ [\text{A}]$$

$$S_{\text{base}} = \frac{\text{Base voltage} \times \text{base current}}{1000} = \frac{600 \times 6}{1000} = 3.6\ [\text{kVA}] = 3600\ [\text{VA}]$$

Example 6.2

A 115 [kV] single-phase power system has a base voltage value of 100 [kV] and a base current value of 400 [A]. Determine the base impedance and per-unit voltage of the power system.

Solution

We have

$$Z_{\text{base}} = \text{Base impedance} = \frac{\text{base voltage}}{\text{base current}} = \frac{100 \times 10^3}{400} = 250\ [\Omega]$$

$$V_{\text{pu}} = \text{Per-unit voltage} = \frac{\text{actual voltage}}{\text{base voltage}} = \frac{115 \times 10^3}{100 \times 10^3} = 1.15\ [\text{pu}]$$

Example 6.3

A three-phase wye-connected synchronous generator has a three-phase apparent power rating of 100 [MVA], a line-to-line voltage rating of 30 [kV], and a per-phase synchronous reactance value of 12 [Ω]. Determine the generator per-unit reactance using rated machine apparent power and voltage as base values. Also determine the generator per-unit reactance value on 120 [MVA], 30 [kV] base.

Solution

$$X_{\text{base}} = \text{Base reactance} = \frac{{V_{\text{base},\,LL}}^2}{S_{\text{base},\,3\phi}} = \frac{\left(30 \times 10^3\right)^2}{100 \times 10^6} = 9\ [\Omega]$$

$$X_{\text{pu}} = \text{Per-unit reactance} = \frac{\text{actual reactance}}{\text{base reactance}} = \frac{12}{9} = 1.33\ [\text{pu}]$$

The generator reactance on the new 120 [MVA], 30 [kV] base is

$$X_{\text{pu on new base}} = X_{\text{pu on old base}} \times \frac{V_{\text{old base}}^2}{V_{\text{new base}}^2} \times \frac{S_{\text{new base}}}{S_{\text{old base}}}$$

$$= 1.33 \times \left(\frac{30}{30}\right)^2 \times \left(\frac{120}{100}\right) = 1.6\ [\text{pu}]$$

6.4 PER-PHASE, PER-UNIT EQUIVALENT CIRCUITS

We earlier learned that three-phase power systems are analyzed by using their per-phase equivalent circuit to determine system voltages and currents at various locations in one of the phases [6,38,46]. If a three-phase circuit contains delta-connected loads, the loads are converted to equivalent wye connections using delta-wye transformations, and per-phase analysis is then performed. Voltages and currents in other two phases of the system have the same magnitude and a progressive phase difference of 120°. A three-phase AC circuit example in an earlier chapter illustrated the per-phase analysis method.

It was also described earlier in this chapter that the per-unit method of calculations has much usefulness in obtaining the solution of power system problems containing many transformers since different voltage levels disappear and the power system network consisting of various components reduces to a system of impedances.

Real power systems are three-phase systems consisting of numerous delta and wye connections and many transformers. They must be analyzed using their per-phase equivalent circuit since they are three-phase systems. They must also be analyzed using the per-unit method of calculations since they consist of many transformers and therefore many different voltage levels. Therefore, for the analysis of real power systems, their per-phase, per-unit equivalent circuit must be constructed [6,46].

Some observations must be remembered while constructing a per-phase, per-unit equivalent circuit of a power system. Base three-phase apparent power and base line-to-line voltage values are selected at a particular location in a power system. The base apparent power will be the same for the entire power system. The ratio of voltage bases on the two sides of a transformer will be the same as its turns ratio or line-to-line voltage ratio. The per-unit impedance value of a power system component such as a transmission line is obtained by dividing its actual impedance value in Ohms with base impedance value for the corresponding section of the power system in which the component is located. Impedance of a power system component such as a generator provided in per-unit on a base other than that calculated for the section in which the component is located is converted to the correct base using equations described above [6,9,38,46]. The procedure for the construction of a per-phase, per-unit equivalent circuit of a power system is described by means of an example below.

A power system per-phase, per-unit equivalent circuit impedance diagram can be converted to a reactance diagram if the resistance is neglected. Omission of resistance introduces some error; however, the results can be acceptable since the inductive reactance of a power system is generally much larger than its resistance.

Example 6.4

A simple power system is shown in Figure 6.4, which consists of a synchronous generator, a synchronous motor load, and a three-phase load connected by two transformers and a transmission line. The data values for the power system components are shown in Table 6.1.

The resistances of generators, transformers, and motors are negligible. The transmission line has a reactance of 44 [Ω] and its resistance can also be neglected. The three-phase load at bus 4 consumes 25 [MVA], with 0.8 power factor lagging at 9.5 [kV].

Draw a per-phase per-unit equivalent circuit impedance diagram for this simple power system using base apparent power and base line voltage of 100 [MVA] and 20 [kV], respectively, on the primary side of transformer T_1.

Solution

We first determine voltage bases for all sections of the network.

The base voltage V_{base1} on the low-voltage side of T_1 is 20 [kV].

The base voltage V_{base2} on the high-voltage side at bus 2 is

$$V_{\text{base2}} = 20\left(\frac{200}{20}\right) = 200 \text{ [kV]}$$

The base voltage V_{base3} at bus 3 on the high-voltage side of T_2 is also 200 [kV].

The base voltage V_{base4} on the low-voltage side of T_2 at bus 4 is

$$V_{\text{base4}} = 200\left(\frac{10}{200}\right) = 10 \text{ [kV]}$$

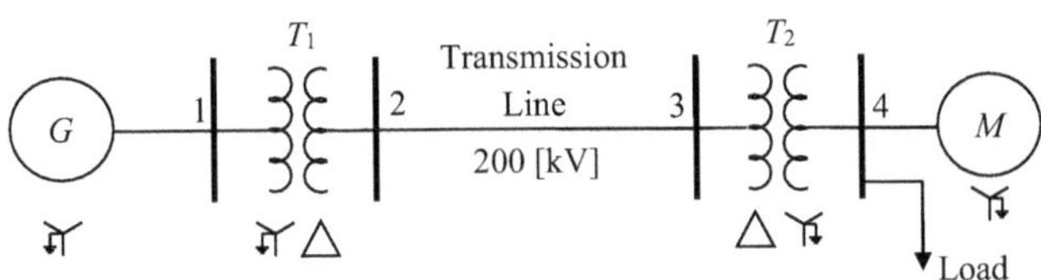

FIGURE 6.4 Example power system.

TABLE 6.1
Data Values of Various Power System Components

Power System Component	Three-Phase Apparent Power Rating (MVA)	Line Voltage Rating (kV)	Reactance Values Based on Equipment Nameplate Rating (pu)
Generator G	100	20	0.15
Transformer T_1	120	20/200	0.12
Transformer T_2	110	200/10	0.08
Motor M	50	9.5	0.16

The voltage bases of generator G and transformers T_1 and T_2 are the same as their rated values. Their per-unit reactances on a chosen MVA base of 100 [MVA] are determined using the following equation described above:

$$X_{\text{pu on new base}} = X_{\text{pu on old base}} \times \frac{S_{\text{new base}}}{S_{\text{old base}}}$$

We therefore have

$$X_G\,(= X_s) = 0.15\left(\frac{100}{100}\right) = 0.15\ [\text{pu}]$$

$$X_{T_1} = 0.12\left(\frac{100}{120}\right) = 0.10\ [\text{pu}]$$

$$X_{T_2} = 0.08\left(\frac{100}{110}\right) = 0.073\ [\text{pu}]$$

The motor reactance is expressed based on its rating of 50 [MVA] and 9.5 [kV]. The base voltage, however, at bus 4 for the motor is 10 [kV]. We therefore use the following equation described earlier to determine the motor reactance for the base values of 100 [MVA] and 10 [kV]:

$$X_{\text{pu on new base}} = X_{\text{pu on old base}} \times \frac{V_{\text{old base}}^{\;2}}{V_{\text{new base}}^{\;2}} \times \frac{S_{\text{new base}}}{S_{\text{old base}}}$$

We thus have

$$X_M = 0.16 \times \left(\frac{9.5}{10}\right)^2 \times \left(\frac{100}{50}\right) = 0.29\ [\text{pu}]$$

The impedance base for the transmission line with a base line voltage of 200 [kV] and a base three-phase apparent power of 100 [MVA] is

$$Z_{\text{base2}} = \frac{\left(200 \times 10^3\right)^2}{100 \times 10^6} = 400\ \Omega$$

The per-unit reactance X_{TL} of the transmission line is

$$X_{\text{TL}} = \left(\frac{44}{400}\right) = 0.11\ [\text{pu}]$$

The three-phase apparent power of load at 0.8 power factor lagging at a voltage level of 9.5 [kV] is

$$\mathbf{S}_{\text{Load},3\phi} = 25\angle 36.87^\circ\ [\text{MVA}]$$

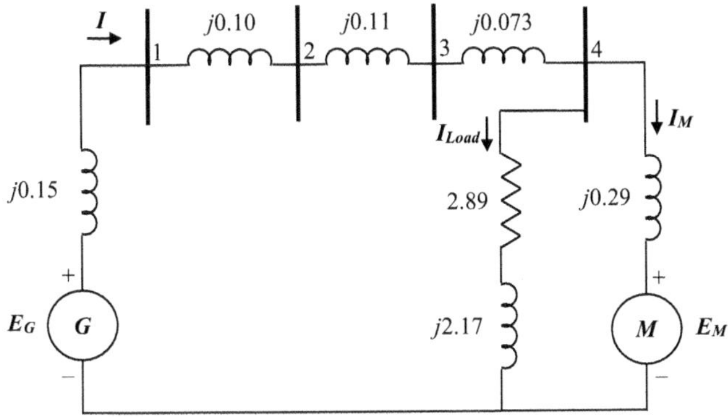

FIGURE 6.5 Example per-phase per-unit impedance diagram.

The per-phase load impedance in Ohms, therefore, is

$$\boldsymbol{Z}_{\text{Load}} = \frac{\left(9.5\times10^{3}\right)^{2}}{(25\angle-36.87°)\times10^{6}} = (2.89 + j2.17)\ [\Omega]$$

The base impedance for the load connected to bus 4 with a base line voltage of 10 [kV] and a base three-phase apparent power of 100 [MVA] is

$$Z_{\text{base4}} = \frac{\left(10\times10^{3}\right)^{2}}{100\times10^{6}} = 1\ [\Omega]$$

The load impedance, therefore, in per-unit is

$$\boldsymbol{Z}_{\text{Load}} = \frac{2.89 + j2.17}{1} = (2.89 + j2.17)\ [\text{pu}]$$

The equivalent per-phase per-unit impedance diagram is shown in Figure 6.5.

Example 6.5

In the previous example, determine the generator and motor internally generated voltages $\boldsymbol{E}_G$ and $\boldsymbol{E}_M$, respectively, if the motor is consuming full rated apparent power of 50 [MVA] and has a power factor of 0.9 lagging at a voltage of 9.5 [kV].

Solution

The voltage at bus 4 to which the motor is connected is assumed as a reference. Per-unit voltage at bus 4, therefore, is

$$\boldsymbol{V}_4 = \left(\frac{9.5\angle0°}{10}\right) = 0.95\angle0°\ [\text{pu}]$$

Motor per-unit apparent power at a power factor of 0.9 lagging

$$S_M = \left(\frac{50\angle 25.84^\circ}{100}\right) = 0.50\angle 25.84^\circ\ [\text{pu}]$$

We therefore have

$$I_M = \frac{S_M^*}{V_4^*} = \left(\frac{0.50\angle -25.84^\circ}{0.95\angle 0^\circ}\right) = 0.526\angle -25.84^\circ\ [\text{pu}]$$

$$I_{\text{Load}} = \frac{V_4}{Z_{\text{Load}}} = \left(\frac{0.95\angle 0^\circ}{2.89 + j2.17}\right) = 0.263\angle -36.90^\circ\ [\text{pu}]$$

Total bus 4 current

$$I = I_M + I_{\text{Load}} = 0.786\angle -29.52^\circ\ [\text{pu}]$$

Total impedance between bus 1 and bus 4

$$Z_{1-4} = j0.283\ [\text{pu}]$$

Bus 1 voltage

$$\begin{aligned} V_1 = V_4 + Z_{1-4} I &= 0.95\angle 0^\circ + j0.283 \times (0.786\angle -29.52^\circ) \\ &= 1.077\angle 10.35^\circ\ [\text{pu}] \\ &= 21.54\angle 10.35^\circ\ [\text{kV}] \end{aligned}$$

The generator internally generated voltage is

$$\begin{aligned} E_G = V_1 + Z_G I &= (1.077\angle 10.35^\circ) + j0.15 \times (0.786\angle -29.52^\circ) \\ &= 1.156\angle 14.84^\circ\ [\text{pu}] \\ &= 23.12\angle 14.84^\circ\ [\text{kV}] \end{aligned}$$

The motor internally generated voltage is

$$\begin{aligned} E_M = V_4 - Z_M I_M &= 0.95\angle 0^\circ - j0.29 \times (0.526\angle -25.84^\circ) \\ &= 0.894\angle -8.84^\circ\ [\text{pu}] \\ &= 8.94\angle -8.84^\circ\ [\text{kV}] \end{aligned}$$

6.5 SUMMARY

Power systems are usually drawn in one-line diagram form in which three-phase transmission line is represented by a single line and standard symbols are used to represent various power system components. One-line diagrams also indicate the type of equipment connections such as delta and wye connections and the locations where power system components are connected to ground. They also many times have the relevant data pertaining to various components such as equipment power ratings, voltages, resistances, and reactances. The amount of information shown on the one-line diagram depends on the purpose such as power flow analysis and fault analysis for which it is drawn.

To solve problems of three-phase power systems containing transformers, the per-phase per-unit analysis is commonly utilized. The per-phase analysis is used since system voltages and currents at various locations in the three phases of a power system have equal magnitude and progressive phase difference of 120°. The per-unit analysis is used since its application removes voltage-level changes occurring due to the presence of many transformers in power systems.

REVIEW QUESTIONS

6.1 Describe why one-line diagrams are used to represent power systems.

6.2 Describe the different types of symbols used in power system one-line diagrams.

6.3 Describe how the application of per-unit system simplifies the analysis of power system problems containing transformers.

6.4 Derive the per-unit system voltage, current, and impedance relationships for single-phase power systems.

6.5 Derive the per-unit system voltage, current, and impedance relationships for three-phase power systems.

6.6 Describe how the per-unit impedance of a generator given in the machine base can be converted to the power system base.

6.7 Describe the per-phase per-unit equivalent circuit method used for solving power system problems.

6.8 Describe how the per-unit impedance diagram of a power system is converted to a per-unit reactance diagram.

7 Power System Operation

7.1 INTRODUCTION

The power stations generate electricity in large amounts, and since the customers are located over a large geographical area, the generated electrical energy must be transmitted using an electrical power network.

Electric generators used for power generation in power stations have become larger, and they can have power ratings of 2000 [MW] or more. The electric power network must be capable of meeting the needs of all customers while transmitting power generated in the power plants. Electric power networks transmit large quantities of electric power through the grid or transmission lines. A sub-transmission network system receives power from the grid in smaller quantities. Finally, individual customers are served using the distribution network.

The following sections describe control of the power system frequency and the bus voltage profile, which are the most important tasks in the proper operation of the power system. Some other fundamentals related to the control of real and reactive power flows in transmission lines and transmission losses are also described below.

7.2 STRUCTURE OF THE ELECTRIC POWER NETWORK

Transformers at the transmission substations raise the voltage generated by generators, which is typically in the 11–30 [kV] range, to the primary transmission voltage. At transmission voltage levels typically between 230 and 765 [kV], large amounts of power are transmitted from the generating stations to the load-center substations located close to the consumers in the USA. The power grid is a network made up of these very high-voltage transmission lines. Most of the large and efficient power plants connect directly to this network through power transformers. A lower voltage sub-transmission network, which operates at voltage levels between 69 and 138 [kV], is fed by this grid in turn. The distribution system operates at voltage levels between 4 and 34.5 [kV] and electricity is supplied to final consumer feeders at voltage level of 208 [V] three-phase, with 120 [V] between phase and neutral. Figure 7.1 shows the different voltage level networks that make up the power system. Simplified representation of a typical portion of the power transmission system or the power grid is shown in Figure 7.2. The generated voltage is raised to transmission level voltage using transformer at the transmission substation and supplied to the power grid through the circuit breaker. The power grid has a loop structure, which is different from radial structure of the distribution system. Shunt capacitors are connected to non-generator buses to maintain voltage levels of these buses [12,48]. A transformer is shown to connect 345 and 500 [kV] portions of the power transmission system in this simplified representation (in reality, a slightly different design of transformer without isolation between primary and secondary windings is more commonly utilized for the purpose of connecting such transmission lines).

DOI: 10.1201/9781003432340-7

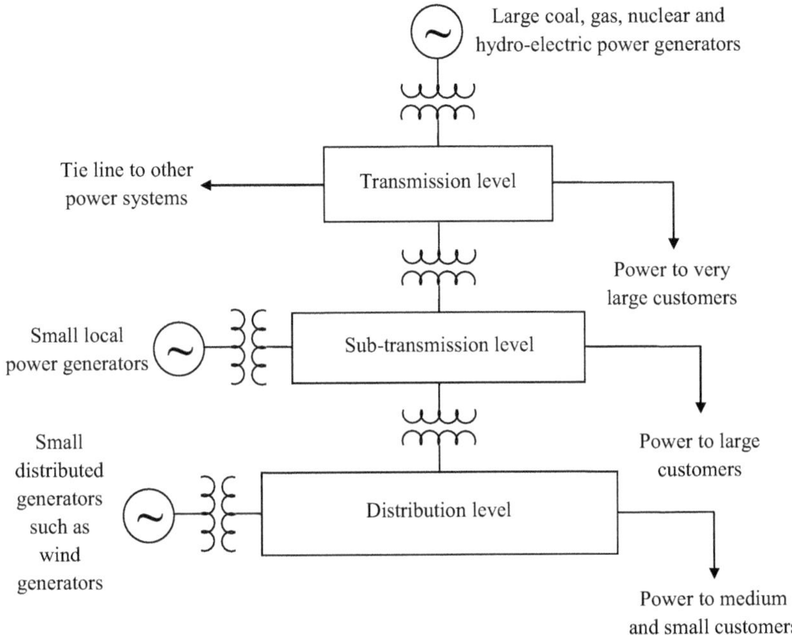

FIGURE 7.1 Schematic diagram of the constituent networks of a power supply system [48].

Many generating units are large and efficient, and they are operated 24 hours a day due to economic reasons. These high-efficiency stations are interconnected as described above, and they supply power to the main base load. Some extra amount of generating capacity, referred to as the spinning reserve, is needed to handle abrupt increases in load. It consists of partially loaded generators that are synchronized with the system and operating to supply power instantaneously. In the entire USA, the electricity supplies are synchronized, and 60 [Hz] is the common power system frequency. Interconnection provides alternative passageways between generators and bulk supply locations supplying the distribution systems for improved reliability [12,48].

7.3 POWER SYSTEM LOAD VARIATION

A typical city's daily load demand curve as a function of time was shown earlier in Figure 2.8. Power demand changes smoothly and predictably from hour to hour. Using appropriate seasonal daily and weekly load demand curves, the power company can forecast the load with adequate accuracy from one day to the next, allowing it to plan for generation properly. The description here about real power loads applies to reactive loads too. Reactive power is typically consumed in proportion to real power by all electric loads [38,48].

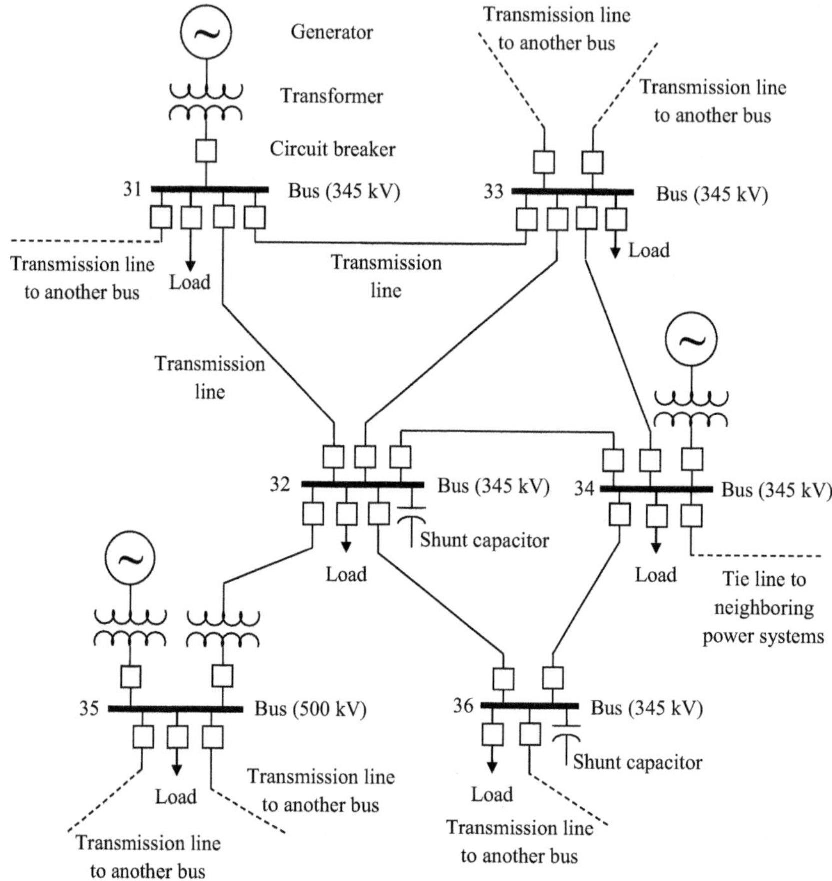

FIGURE 7.2 Simplified representation of a portion of the power grid [12].

When observed on hourly basis, load demand curves have a generally smooth and predictable pattern, as described in Chapter 2. However, if they are observed on the second-by-second basis or minute-by-minute basis, they have different appearances.

A closer view of the load demand curve over a 30-minute period is shown in Figure 7.3. The average demand is 950 [MW], with a random variation of about 5 [MW] above or below the average value. Because it is impossible to determine instantaneous consumer demand with proper accuracy, the noise component is completely unpredictable. There is also a clear relationship between power system frequency and load. The frequency will decrease as a result of the use of kinetic energy of the moving masses of the power generating system to supply the sudden increase in load if the generator power output is maintained constant [12,38,48].

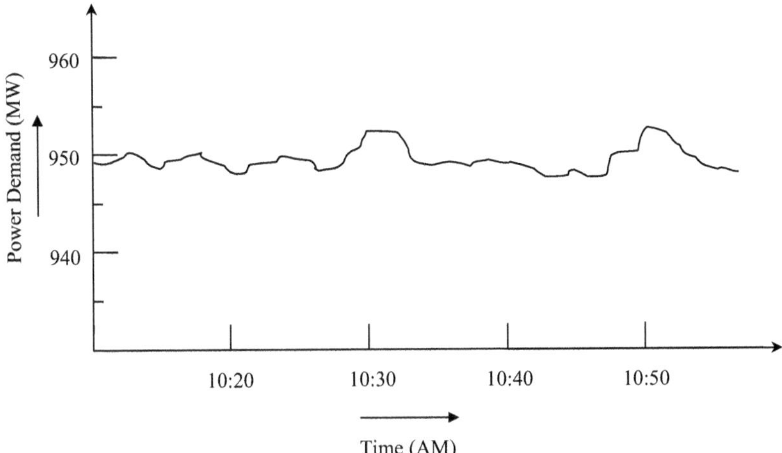

FIGURE 7.3 Power demand curve over a 30-minute interval [12].

7.4 CONTROL OF POWER SYSTEM FREQUENCY

The real power balance across the entire power network and the power system frequency are closely related [11,12,48]. The power system generators operate synchronously and produce the energy needed to supply power to all loads at any instant, plus any real transmission losses, under normal operating conditions. The real transmission losses are typically only a small percentage of the overall power generated.

Since electrical energy is not stored anywhere in electric form in the power system and is carried at virtually the speed of light, the rate at which it is produced and used at any one time must be equal. The average amount of storage occurring in reactive elements, such as coils and capacitors twice per electrical cycle, is zero.

In case of real power imbalance, the corresponding difference will be supplied from or absorbed by the generator rotor kinetic energy storage. Since the rotor kinetic energy depends on rotor speed, a power imbalance will result in speed variation and frequency deviation.

Automatic regulation is utilized to regulate the frequency in all modern power systems. The functioning of the automatic load-frequency control (ALFC) system is depicted in Figure 7.4. The power system frequency f is sensed by a frequency sensor-comparator, and it is compared to reference frequency f_{ref} (60 [Hz]). The resulting frequency error signal is

$$\Delta f = f_{ref} - f \text{ [Hz]} \tag{7.1}$$

The error signal is sent to the steam valve controller for changing the steam turbine control valve's setting. In case of a positive error signal, which denotes low system frequency, the steam valve will be given a raise command, causing the generator power output P_G to increase. If the error signal is negative, a lower command will be given to the steam valve to reduce the turbine steam input and decrease the generator's power output P_G.

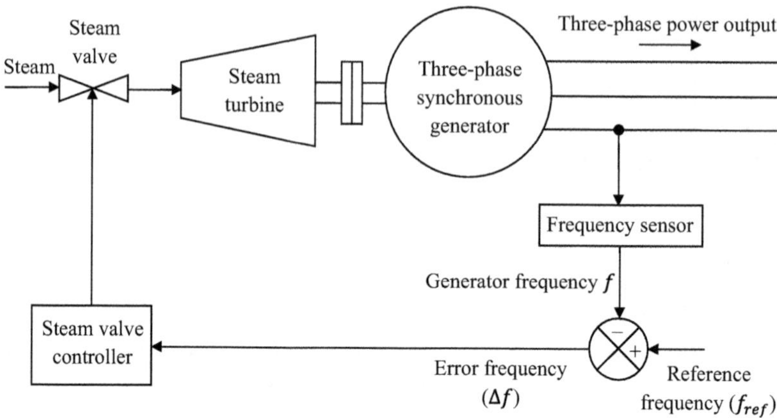

FIGURE 7.4 Automatic load-frequency control system.

7.5 CONTROL OF BUS VOLTAGE

It is necessary to maintain stable voltage magnitudes at each bus during power system operation, since all electric loads are voltage rated, and as a result, are built to function at a specific voltage [11,12,48]. For instance, an incandescent light bulb's light output varies depending on the system voltage. Variations in light flux due to voltage fluctuation can result in customer annoyance.

The automatic excitation control (AEC) of individual generators is used to regulate the power system bus voltages. Since the generator emf is proportional to the rotor field current, an AEC system can be developed as depicted in Figure 7.5.

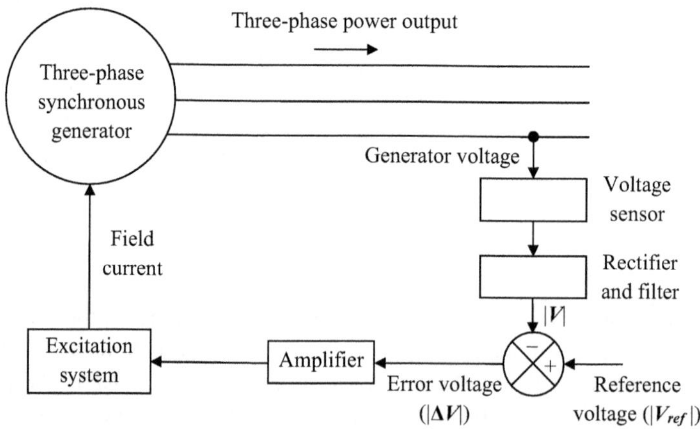

FIGURE 7.5 Automatic excitation control (AEC) loop for line voltage profile control.

The generator output voltage is sensed, rectified and filtered, and the resulting voltage $|V|$ is compared with a reference voltage $|V_{\text{ref}}|$ to produce the following voltage error signal $|\Delta V|$

$$|\Delta V| = |V_{\text{ref}}| - |V| \ [\text{V}] \tag{7.2}$$

The amplified error voltage is then sent to the field-current source as an actuating signal to regulate the field current, thereby correcting the voltage variation and reducing the error voltage.

The voltage level of a bus is also strongly related to the reactive power injection at the bus. Added reactive power results in a rise in bus voltage.

The two main control mechanisms for the synchronous generator, therefore, are the AEC loop and the ALFC loop, as described above. These two loops do not interact with each other and therefore the working of one loop has negligible effect on that of the other. Since the AEC loop operation includes electrical variables only, it responds much more quickly than the ALFC loop, which involves both electrical variables and mechanical variables like steam turbine mechanical inertia and steam valves [11,12].

7.6 CONTROL OF REAL AND REACTIVE POWER FLOWS IN TRANSMISSION LINES

In this section, we first examine how transmission line real power flow can be controlled both in magnitude and direction [12,38]. We assume the transmission line to be lossless with $R=0$ since it simplifies the explanation. The simplified transmission line representation is shown in Figure 7.6, which has the sending end current equal to the receiving end current.

We have the following equations for complex power at each end of the transmission line:

$$S_1 = P_1 + jQ_1 = V_1 I^* \ [\text{VA}] \tag{7.3}$$

$$S_2 = P_2 + jQ_2 = V_2 I^* \ [\text{VA}] \tag{7.4}$$

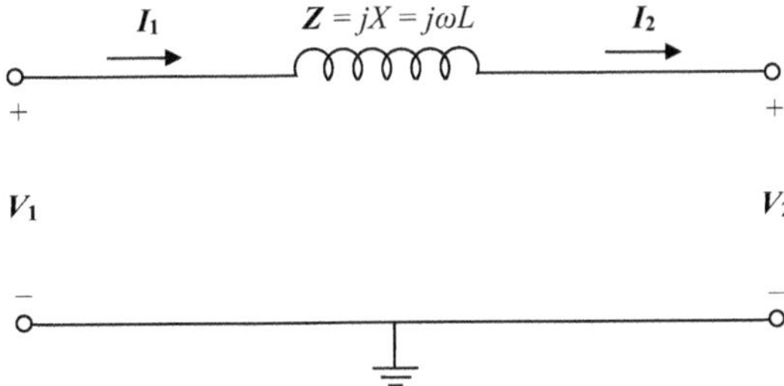

FIGURE 7.6 Per-phase representation of short transmission line [38].

The impedance of the transmission line and the line current are

$$\boldsymbol{Z} = jX = j\omega L \tag{7.5}$$

$$\boldsymbol{I} = \frac{\boldsymbol{V}_1 - \boldsymbol{V}_2}{\boldsymbol{Z}} \ [\text{A}] \tag{7.6}$$

We substitute $\boldsymbol{I}$ from equation (7.6) in $\boldsymbol{S}_1$ equation (7.3) and obtain

$$\begin{aligned} P_1 + jQ_1 &= \boldsymbol{V}_1 \frac{\boldsymbol{V}_1^* - \boldsymbol{V}_2^*}{\boldsymbol{Z}^*} = \frac{|\boldsymbol{V}_1|^2 - |\boldsymbol{V}_1||\boldsymbol{V}_2| e^{j\delta}}{-jX} \\ &= \frac{|\boldsymbol{V}_1||\boldsymbol{V}_2|}{X} \sin\delta + j\frac{|\boldsymbol{V}_1|^2 - |\boldsymbol{V}_1||\boldsymbol{V}_2|\cos\delta}{X} \ [\text{VA}] \end{aligned} \tag{7.7}$$

where

$$\delta = \angle \boldsymbol{V}_1 - \angle \boldsymbol{V}_2 \tag{7.8}$$

Separation of real and imaginary parts in equation (7.7) yields

$$P_1 = \frac{|\boldsymbol{V}_1||\boldsymbol{V}_2|}{X} \sin\delta \ [\text{W}] \tag{7.9}$$

$$Q_1 = \frac{|\boldsymbol{V}_1|^2 - |\boldsymbol{V}_1||\boldsymbol{V}_2|\cos\delta}{X} \ [\text{VAR}] \tag{7.10}$$

We can similarly derive the following equations below:

$$P_2 = \frac{|\boldsymbol{V}_1||\boldsymbol{V}_2|}{X} \sin\delta \ [\text{W}] \tag{7.11}$$

$$Q_2 = \frac{|\boldsymbol{V}_2||\boldsymbol{V}_1|\cos\delta - |\boldsymbol{V}_2|^2}{X} \ [\text{VAR}] \tag{7.12}$$

From the above equations, it is possible to determine the power-carrying limits of the transmission line. The phase angle between $\boldsymbol{V}_1$ and $\boldsymbol{V}_2$ is referred to as the power angle δ.

To obtain per-phase values of powers in the above equations, per-phase values of the voltages and reactance must be provided. Since the transmission line has zero resistance, real power at the sending end is equal to that at the receiving end, which means $P_1 = P_2 = P$. As the transmission line reactance X is fixed and voltage magnitude values $|\boldsymbol{V}_1|$ and $|\boldsymbol{V}_2|$ are kept constant, P in equations (7.9) and (7.11) is a function of power angle δ only.

The transmitted real power P will increase as power angle δ is increased with $|V_1|$ leading $|V_2|$, and it will have the following maximum value for $\delta = 90°$:

$$P_{max} = \frac{|V_1||V_2|}{X} \text{ [W per phase]} \tag{7.13}$$

The transmitted real power P will decrease instead of increasing if we attempt to increase it by further raising δ. If V_2 leads V_1 and δ increase is negative, real power flows from the receiving end to the sending end or from right to left. In the case of AC power transmission, real power flows from the side with a leading voltage phase angle to the side with a lagging voltage phase angle.

We next learn how reactive power flow in a transmission can be controlled in terms of magnitude and direction. We can rewrite equations (7.10) and (7.12) as follows:

$$Q_1 = \frac{|V_1|}{X}\left(|V_1| - |V_2|\cos\delta\right) \text{ [VAR]} \tag{7.14}$$

$$Q_2 = \frac{|V_2|}{X}\left(|V_1|\cos\delta - |V_2|\right) \text{ [VAR]} \tag{7.15}$$

The value of $\cos\delta$ is typically close to 1 for power system operating conditions. As a result, the terms enclosed in parenthesis in both Q_1 and Q_2 equations are roughly proportional to $\left(|V_1| - |V_2|\right)$. We therefore find that reactive power moves from the higher voltage end of the transmission line to its lower voltage end. The difference between the voltage magnitudes at the two ends of the transmission line determines the value of the reactive power flow.

Example 7.1

A 75-km long transmission line has negligible resistance and a per-phase inductive reactance of 50 [Ω]. It has a sending end line-to-line voltage of 345 [kV] and a receiving end line-to-line voltage of 330 [kV]. Determine the maximum power which can be transmitted along this transmission line.

Solution

Maximum transmitted three-phase real power P_{max} is

$$P_{max} = 3 \times \frac{|V_1||V_2|}{X}$$

In the above equation, $|V_1|$ and $|V_2|$ are phase voltage magnitudes at the sending end and at the receiving end, respectively, and X is the transmission line per-phase inductive reactance.

We therefore have

$$P_{max} = 3\times\frac{\left(345\times10^3/\sqrt{3}\right)\times\left(330\times10^3/\sqrt{3}\right)}{50} = 2277\times10^6\ [\text{W } 3\phi]$$

$$= 2277\ [\text{MW } 3\phi]$$

$$\approx 2.28\ [\text{GW } 3\phi]$$

Example 7.2

Determine the reactive power flow in the lossless transmission line of the previous example. Assume power angle δ to be 10°.

Solution

The phase voltages at the sending end and the receiving end are

$$|V_1| = \frac{345}{\sqrt{3}} = 199.19\ [\text{kV}]$$

$$|V_2| = \frac{330}{\sqrt{3}} = 190.53\ [\text{kV}]$$

From equations (7.14) and (7.15), we have

$$Q_1 = \frac{|V_1|}{X}\left(|V_1| - |V_2|\cos\delta\right)$$

$$= \frac{199.19\times10^3}{50}\left(199.19\times10^3 - 190.53\times10^3\times\cos10°\right)$$

$$= 46.03\ [\text{MVAR per phase}]$$

$$= 138.09\ [\text{MVAR } 3\phi]$$

$$Q_2 = \frac{|V_2|}{X}\left(|V_1|\cos\delta - |V_2|\right)$$

$$= \frac{190.53\times10^3}{50}\left(199.19\times10^3\times\cos10° - 190.53\times10^3\right)$$

$$= 21.47\times10^6\ [\text{VAR per phase}]$$

$$= 21.47\ [\text{MVAR per phase}]$$

$$= 64.41\ [\text{MVAR } 3\phi]$$

Reactive power consumed in the inductive reactance of the transmission line is 73.68 $[\text{MVAR } 3\phi]$, which is the difference between Q_1 and Q_2.

7.7 TRANSMISSION LINE POWER LOSSES

Equations (7.9)–(7.12) for P and Q were obtained after assuming that the transmission line had zero resistance and therefore transmission line real power loss was zero [38]. Series resistance in a real transmission line will, however, result in Ohmic power loss P_Ω, which can be determined using the following equation:

$$P_\Omega = |I|^2 R \left[\text{W per phase}\right] \tag{7.16}$$

Since transmission line voltage, current, and power values differ at different locations, we assume V, I, P, and Q to be values measured at the middle of the transmission line in the equations below. We therefore have

$$P + jQ = \boldsymbol{V}\boldsymbol{I}^* \;[\text{VA}] \tag{7.17}$$

$$\boldsymbol{I}^* = \frac{P + jQ}{\boldsymbol{V}} \;[\text{A}]$$

$$\boldsymbol{I} = \frac{P - jQ}{\boldsymbol{V}^*} \;[\text{A}] \tag{7.18}$$

$$\boldsymbol{I} \times \boldsymbol{I}^* = |\boldsymbol{I}|^2 = \frac{P - jQ}{\boldsymbol{V}^*} \times \frac{P + jQ}{\boldsymbol{V}} = \frac{P^2 + Q^2}{|\boldsymbol{V}|^2} \tag{7.19}$$

$$P_\Omega = |\boldsymbol{I}|^2 R = \frac{P^2 + Q^2}{|\boldsymbol{V}|^2} R \left[\text{W per phase}\right] \tag{7.20}$$

From the above P_Ω equation, we can conclude that both real and reactive line powers are equally responsible for the Ohmic power loss in the transmission line. It is therefore important to reduce both real and reactive power flows in order to minimize Ohmic power loss during transmission [38]. In an actual power system, it is generally achieved by producing the reactive power at the bus where it is required using shunt capacitors or an overexcited synchronous generator.

Example 7.3

The average value of P, Q measured at the middle location of a 60-km long transmission line having per-phase resistance of 10 $[\Omega]$ are 180 $\left[\text{MW } 3\phi\right]$ and 90 $\left[\text{MVAR } 3\phi\right]$. The line-to-line voltage at the middle location of the transmission line is 230 [kV]. Determine the Ohmic power loss in the transmission line.

Solution

The phase voltage at the middle location of the transmission line is

$$|V| = \frac{230}{\sqrt{3}} = 132.79 \ [\text{kV}]$$

The per-phase values of P and Q are 60 [MW] and 30 [MVAR], respectively.
We have from equation (7.20)

$$P_{\Omega} = \frac{P^2 + Q^2}{|V|^2} R$$

$$= \frac{\left(60 \times 10^6\right)^2 + \left(30 \times 10^6\right)^2}{\left(132.79 \times 10^3\right)^2} \times 10$$

$$= 2.55 \times 10^6 \ [\text{W per phase}]$$

$$= 2.55 \ [\text{MW per phase}]$$

$$= 7.65 \ [\text{MW } 3\phi]$$

7.8 SUMMARY

An electric power network links power generation units to individual customers. It must be able to transmit the required amount of power to all its customers, whether large or small, at all times.

Power generation at a constant frequency is a main requirement for proper operation of the power system. An ALFC system maintains constant power system frequency by maintaining the real power balance within the power system at all times. Variation in power system frequency is caused due to imbalance between generation and consumption of real power in the power system.

Maintaining proper voltage profile in the entire power system is another primary requirement for the proper operation of the power system. It is accomplished by AEC of the individual generators, switched shunt capacitors and reactors, and tap changing of transformers.

The Ohmic power loss in the transmission line is caused by the flow of both real and reactive line powers. The transmission losses can therefore be controlled by controlling the real and reactive power flows in the transmission network.

REVIEW QUESTIONS

7.1 What are the main requirements of an electric power network?
7.2 Describe the general structure of the electric power network.
7.3 What are the main tasks in the proper operation of the power system?
7.4 Describe the automatic excitation control utilized for bus voltage control in power systems.

7.5 Describe the automatic load-frequency control utilized for the control of power system frequency.
7.6 Describe how the flow of real power is controlled in a power system.
7.7 Describe how the flow of reactive power is controlled in a power system.
7.8 Describe how the real power loss in a transmission line depends on the flow of real and reactive line powers.

8 Power Flow Analysis

8.1 INTRODUCTION

Power flow analysis deals with the determination of voltage magnitude and angle at each bus in a power system under balanced three-phase steady-state conditions. The values of real and reactive power flows in all transmission lines and power system losses are also obtained from the analysis. The principles underlying power flow analysis are simple; however, such an analysis of a real power system can only be carried out with the help of a digital computer.

In power flow analysis, a suitable mathematical network model is developed to describe the relationships between voltages and powers in an interconnected power system. The power and voltage constraints applicable to the different buses in the power system network are then stated. Subsequently, required numerical computations are carried out using an iterative procedure on the mathematical model developed as power flow equations subject to the aforementioned constraints to determine the magnitudes and phase angles of all bus voltages. The actual power flows and power losses in all transmission lines can be easily obtained after all bus voltages have been determined.

This chapter describes power flow analysis in detail. The control of power flow in electrical power networks and the economic operation of power systems are also briefly described.

8.2 DERIVATION OF POWER FLOW EQUATIONS

8.2.1 Power Flow Equations for Three-Bus System

In power flow analysis formulation, we develop the mathematical network model of the simple three-bus system shown in Figure 8.1 [9,11,38]. The equations derived for this simple system are later extended to power flow analysis of power systems with any number of buses.

Buses 1, 2, and 3 have voltages $\boldsymbol{V}_1$, $\boldsymbol{V}_2$, and $\boldsymbol{V}_3$, respectively. The buses 1, 2, and 3 receive powers $\boldsymbol{S}_{G1}$, $\boldsymbol{S}_{G2}$, and $\boldsymbol{S}_{G3}$, respectively, from generators connected to them. The loads connected to buses 1, 2, and 3 draw powers $\boldsymbol{S}_{D1}$, $\boldsymbol{S}_{D2}$, and $\boldsymbol{S}_{D3}$, respectively, from them. The buses 1 and 2 are connected through a transmission line which has been represented using its equivalent π configuration with series admittance $\boldsymbol{Y}_{s12}$ and two parallel admittances $\boldsymbol{Y}_{p12}$. Similarly, the transmission line connecting buses 2 and 3 is represented using its equivalent π configuration with series admittance $\boldsymbol{Y}_{s23}$ and two parallel admittances $\boldsymbol{Y}_{p23}$. Finally, the transmission line connecting buses 1 and 3 is represented using its equivalent π configuration with series admittance $\boldsymbol{Y}_{s13}$ and two parallel admittances $\boldsymbol{Y}_{p13}$.

DOI: 10.1201/9781003432340-8

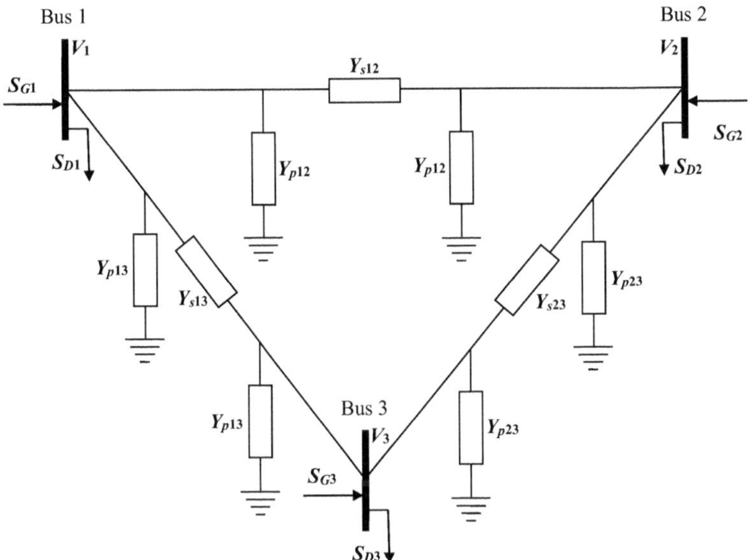

FIGURE 8.1 A simple three-bus example system for power flow analysis.

For each of the buses 1, 2, and 3, the difference between generator power and load power is represented as bus power to simplify the representation in power flow analysis. We therefore have:

$$\begin{aligned} S_1 &= P_1 + jQ_1 = S_{G1} - S_{D1} = (P_{G1} - P_{D1}) + j(Q_{G1} - Q_{D1}) \\ S_2 &= P_2 + jQ_2 = S_{G2} - S_{D2} = (P_{G2} - P_{D2}) + j(Q_{G2} - Q_{D2}) \\ S_3 &= P_3 + jQ_3 = S_{G3} - S_{D3} = (P_{G3} - P_{D3}) + j(Q_{G3} - Q_{D3}) \end{aligned} \tag{8.1}$$

S_1, S_2, and S_3 above are bus powers of buses 1, 2 and 3, respectively.

The bus power S_1 can be represented by equation $S_1 = V_1 I_1^*$, where I_1 is the net current that enters bus 1. This current must be equal to the net current that enters the transmission lines connected to bus 1. The net current entering transmission lines connected to bus 1 has a series admittance components $(V_1 - V_2)Y_{s12}$ and $(V_1 - V_3)Y_{s13}$, and it has shunt admittance components $V_1 Y_{p12}$ and $V_1 Y_{p13}$. We can similarly write series admittance components and shunt admittance components of the net currents entering transmission lines connected to buses 2 and 3.

We therefore have:

$$
\begin{aligned}
\boldsymbol{I}_1 &= \frac{\boldsymbol{S}_1^*}{\boldsymbol{V}_1^*} = (\boldsymbol{V}_1 - \boldsymbol{V}_2)\boldsymbol{Y}_{s12} + (\boldsymbol{V}_1 - \boldsymbol{V}_3)\boldsymbol{Y}_{s13} + \boldsymbol{V}_1\boldsymbol{Y}_{p13} + \boldsymbol{V}_1\boldsymbol{Y}_{p12} \\
\boldsymbol{I}_2 &= \frac{\boldsymbol{S}_2^*}{\boldsymbol{V}_2^*} = (\boldsymbol{V}_2 - \boldsymbol{V}_1)\boldsymbol{Y}_{s12} + (\boldsymbol{V}_2 - \boldsymbol{V}_3)\boldsymbol{Y}_{s23} + \boldsymbol{V}_2\boldsymbol{Y}_{p23} + \boldsymbol{V}_2\boldsymbol{Y}_{p12} \\
\boldsymbol{I}_3 &= \frac{\boldsymbol{S}_3^*}{\boldsymbol{V}_3^*} = (\boldsymbol{V}_3 - \boldsymbol{V}_1)\boldsymbol{Y}_{s13} + (\boldsymbol{V}_3 - \boldsymbol{V}_2)\boldsymbol{Y}_{s23} + \boldsymbol{V}_3\boldsymbol{Y}_{p23} + \boldsymbol{V}_3\boldsymbol{Y}_{p13}
\end{aligned}
\tag{8.2}
$$

The above equations are on a per-phase basis. We next rewrite the above equations as:

$$
\begin{aligned}
\boldsymbol{I}_1 &= \frac{\boldsymbol{S}_1^*}{\boldsymbol{V}_1^*} = \boldsymbol{y}_{11}\boldsymbol{V}_1 + \boldsymbol{y}_{12}\boldsymbol{V}_2 + \boldsymbol{y}_{13}\boldsymbol{V}_3 \\
\boldsymbol{I}_2 &= \frac{\boldsymbol{S}_2^*}{\boldsymbol{V}_2^*} = \boldsymbol{y}_{21}\boldsymbol{V}_1 + \boldsymbol{y}_{22}\boldsymbol{V}_2 + \boldsymbol{y}_{23}\boldsymbol{V}_3 \\
\boldsymbol{I}_3 &= \frac{\boldsymbol{S}_3^*}{\boldsymbol{V}_3^*} = \boldsymbol{y}_{31}\boldsymbol{V}_1 + \boldsymbol{y}_{32}\boldsymbol{V}_2 + \boldsymbol{y}_{33}\boldsymbol{V}_3
\end{aligned}
\tag{8.3}
$$

where

$$
\begin{aligned}
\boldsymbol{y}_{11} &= \boldsymbol{Y}_{p12} + \boldsymbol{Y}_{p13} + \boldsymbol{Y}_{s12} + \boldsymbol{Y}_{s13} \\
\boldsymbol{y}_{22} &= \boldsymbol{Y}_{p12} + \boldsymbol{Y}_{p23} + \boldsymbol{Y}_{s12} + \boldsymbol{Y}_{s23} \\
\boldsymbol{y}_{33} &= \boldsymbol{Y}_{p13} + \boldsymbol{Y}_{p23} + \boldsymbol{Y}_{s13} + \boldsymbol{Y}_{s23} \\
\boldsymbol{y}_{12} &= \boldsymbol{y}_{21} = -\boldsymbol{Y}_{s12} \\
\boldsymbol{y}_{23} &= \boldsymbol{y}_{32} = -\boldsymbol{Y}_{s23} \\
\boldsymbol{y}_{31} &= \boldsymbol{y}_{13} = -\boldsymbol{Y}_{s13}
\end{aligned}
$$

We can write the above set of equations in matrix form as follows:

$$
\boldsymbol{I}_{bus} = \boldsymbol{Y}_{bus}\boldsymbol{V}_{bus}
\tag{8.4}
$$

where

$$
\boldsymbol{I}_{\mathbf{bus}} = \begin{bmatrix} \boldsymbol{I}_1 \\ \boldsymbol{I}_2 \\ \boldsymbol{I}_3 \end{bmatrix}
$$

$$V_{\text{bus}} = \begin{bmatrix} V_1 \\ V_2 \\ V_3 \end{bmatrix}$$

$$Y_{\text{bus}} = \begin{bmatrix} y_{11} & y_{12} & y_{13} \\ y_{21} & y_{22} & y_{23} \\ y_{31} & y_{32} & y_{33} \end{bmatrix}$$

$\boldsymbol{I}_{\text{bus}}$, $\boldsymbol{V}_{\text{bus}}$, and $\boldsymbol{Y}_{\text{bus}}$ above are bus current vector, bus voltage vector, and bus admittance matrix, respectively.

After inverting equation (8.4) above, we obtain:

$$\boldsymbol{V}_{\text{bus}} = \boldsymbol{Z}_{\text{bus}} \boldsymbol{I}_{\text{bus}} \tag{8.5}$$

where

$$\boldsymbol{Z}_{\text{bus}} = \boldsymbol{Y}_{\text{bus}}^{-1} = \begin{bmatrix} z_{11} & z_{12} & z_{13} \\ z_{21} & z_{22} & z_{23} \\ z_{31} & z_{32} & z_{33} \end{bmatrix}$$

$\boldsymbol{Z}_{\text{bus}}$ above is called the bus impedance matrix.

The $\boldsymbol{I}_{\text{bus}}$ and $\boldsymbol{V}_{\text{bus}}$ matrix equations obtained above are sets of complex, linear, simultaneous equations. If, for example, the bus voltages are known, the unknown bus currents can be computed easily using $\boldsymbol{I}_{\text{bus}}$ equation. Similarly, unknown bus voltages can be determined easily using $\boldsymbol{V}_{\text{bus}}$ equation if bus currents are provided.

In real power systems, bus powers and not bus currents are generally known. We therefore obtain from the current equations (8.3) above after some adjustment:

$$\begin{aligned} \boldsymbol{S}_1^* &= P_1 - jQ_1 = \boldsymbol{V}_1^* \boldsymbol{I}_1 = y_{11} V_1 V_1^* + y_{12} V_2 V_1^* + y_{13} V_3 V_1^* \\ \boldsymbol{S}_2^* &= P_2 - jQ_2 = \boldsymbol{V}_2^* \boldsymbol{I}_2 = y_{21} V_1 V_2^* + y_{22} V_2 V_2^* + y_{23} V_3 V_2^* \\ \boldsymbol{S}_3^* &= P_3 - jQ_3 = \boldsymbol{V}_3^* \boldsymbol{I}_3 = y_{31} V_1 V_3^* + y_{32} V_2 V_3^* + y_{33} V_3 V_3^* \end{aligned} \tag{8.6}$$

We can also write the set of equations $\boldsymbol{S}_1^*$, $\boldsymbol{S}_2^*$, and $\boldsymbol{S}_3^*$ obtained above as follows:

$$\begin{aligned} P_1 - jQ_1 &= V_1^* \sum_{k=1}^{3} y_{1k} V_k \\ P_2 - jQ_2 &= V_2^* \sum_{k=1}^{3} y_{2k} V_k \\ P_3 - jQ_3 &= V_3^* \sum_{k=1}^{3} y_{3k} V_k \end{aligned} \tag{8.7}$$

We can further rewrite the above equations more simply as follows:

$$P_i - jQ_i = V_i^* \sum_{k=1}^{3} y_{ik} V_k \quad (i = 1,\ 2,\ 3) \tag{8.8}$$

$$P_i - jQ_i = \sum_{k=1}^{3} |y_{ik}||V_i||V_k| e^{j(\delta_k - \delta_i + \delta_{ik})} \quad (i = 1,\ 2,\ 3) \tag{8.9}$$

In the above equation, δ_i and δ_k are the bus voltage phase angles of buses i and k, respectively, and γ_{ik} is the phase angle of admittance y_{ik}.

From the above set of equations, we can obtain the real and imaginary components:

$$\begin{aligned} P_i = P_{Gi} - P_{Di} &= \sum_{k=1}^{3} |y_{ik}||V_i||V_k| \cos(\delta_k - \delta_i + \gamma_{ik}) \ (i = 1,\ 2,\ 3) \\ Q_i = Q_{Gi} - Q_{Di} &= -\sum_{k=1}^{3} |y_{ik}||V_i||V_k| \sin(\delta_k - \delta_i + \gamma_{ik}) \ (i = 1,\ 2,\ 3) \end{aligned} \tag{8.10}$$

The above power flow equations are nonlinear and they are solved (after slight form adjustment as described below for n-bus system) using numerical iterative procedure with the help of digital computers since it is generally not possible to obtain analytical solutions for them. Ignoring fixed transmission line parameters, the six power flow equations for the three-bus system have 18 variables. The power flow equation variables are P_{G1}, P_{G2}, P_{G3}, Q_{G1}, Q_{G2}, Q_{G3}, P_{D1}, P_{D2}, P_{D3}, Q_{D1}, Q_{D2}, Q_{D3}, $|V_1|$, $|V_2|$, $|V_3|$, δ_1, δ_2, and δ_3. The number of unknowns must be reduced to six by fixing some of the variables in order to solve the equations. We therefore have to specify 12 variables and then the power flow equations can be solved for remaining six variables.

8.2.2 Power Flow Equations for n-Bus System

We next extend the above description for a three-bus system to the general case of an n-bus system [9,11,38]. There are following $6n$ variables for such an n-bus system: n real power generations P_{Gi}, n reactive power generations Q_{Gi}, n real power demands P_{Di}, n reactive power demands Q_{Di}, n-bus voltage magnitudes $|V_i|$, and n-bus voltage phase angles δ_i.

We derive the power flow equation for the n-bus case. The bus power S_i for the general bus i is:

$$S_i = S_{Gi} - S_{Di} = (P_{Gi} - P_{Di}) + j(Q_{Gi} - Q_{Di}) \tag{8.11}$$

S_{Gi} in the above equation is the generator power supplied to bus i and S_{Di} is the load power drawn from bus i for the n-bus case.

Bus i is connected to other buses in the power system network by means of transmission lines. There can be at most $(n-1)$ transmission lines connecting bus i to remaining $(n-1)$ buses in the power system network. We represent transmission line connecting bus i to bus k using its equivalent π configuration with series admittance $\boldsymbol{Y}_{sik}$ and two parallel shunt admittances $\boldsymbol{Y}_{pik}$. If there is no transmission connecting bus i to any particular bus, the corresponding admittance is simply zero. Similar to equation (8.2) for the three-bus system described above, we can write the following current balance equation for the n-bus case:

$$\begin{aligned} \boldsymbol{I}_i = \frac{\boldsymbol{S}_i^*}{\boldsymbol{V}_i^*} &= \boldsymbol{V}_i \sum_{\substack{k=1 \\ k \neq i}}^{n} \boldsymbol{Y}_{pik} + \sum_{\substack{k=1 \\ k \neq i}}^{n} \boldsymbol{Y}_{sik} \left(\boldsymbol{V}_i - \boldsymbol{V}_k \right) \\ &= \boldsymbol{V}_i \sum_{\substack{k=1 \\ k \neq i}}^{n} \left(\boldsymbol{Y}_{pik} + \boldsymbol{Y}_{sik} \right) + \sum_{\substack{k=1 \\ k \neq i}}^{n} \left(-\boldsymbol{Y}_{sik} \right) \boldsymbol{V}_k \quad \left(\boldsymbol{i} = 1, 2, \ldots, \boldsymbol{n} \right) \end{aligned} \tag{8.12}$$

We can also rewrite the above equations for the n-bus system as follows:

$$\begin{aligned} \boldsymbol{I}_i &= \frac{\boldsymbol{S}_i^*}{\boldsymbol{V}_i^*} = \boldsymbol{y}_{i1}\boldsymbol{V}_1 + \boldsymbol{y}_{i2}\boldsymbol{V}_2 + \ldots + \boldsymbol{y}_{ii}\boldsymbol{V}_i + \ldots + \boldsymbol{y}_{in}\boldsymbol{V}_n \quad \left(i = 1, 2, \ldots, n \right) \\ \boldsymbol{I}_i &= \sum_{k=1}^{n} \boldsymbol{y}_{ik}\boldsymbol{V}_k \end{aligned} \tag{8.13}$$

where

$$\boldsymbol{y}_{ii} = \sum_{\substack{k=1 \\ k \neq i}}^{n} \left(\boldsymbol{Y}_{pik} + \boldsymbol{Y}_{sik} \right)$$

$$\boldsymbol{y}_{ik} = \boldsymbol{y}_{ki} = -\boldsymbol{Y}_{sik}$$

For the n-bus power system, we obtain the following equations in vector form for bus currents and voltages:

$$\boldsymbol{I}_{bus} = \boldsymbol{Y}_{bus}\boldsymbol{V}_{bus}$$

$$\boldsymbol{V}_{bus} = \boldsymbol{Z}_{bus}\boldsymbol{I}_{bus}$$

where

$$\boldsymbol{I}_{\mathbf{bus}} = \begin{bmatrix} \boldsymbol{I}_1 \\ \vdots \\ \boldsymbol{I}_n \end{bmatrix}$$

$$V_{\text{bus}} = \begin{bmatrix} V_1 \\ \vdots \\ V_n \end{bmatrix}$$

$$Y_{\text{bus}} = \begin{bmatrix} y_{11} & \cdots & y_{1n} \\ \vdots & \ddots & \vdots \\ y_{n1} & \cdots & y_{nn} \end{bmatrix}$$

$$Z_{\text{bus}} = Y_{\text{bus}}^{-1} = \begin{bmatrix} z_{11} & \cdots & z_{1n} \\ \vdots & \ddots & \vdots \\ z_{n1} & \cdots & z_{nn} \end{bmatrix}$$

The n-dimensional bus current and bus voltage vectors above are n-dimensional and bus admittance and bus impedance matrices above are $n \times n$ dimensional.

As described earlier for the three-bus case, bus powers and not bus currents are generally known. We therefore obtain from the current equations (8.13) above after some adjustment:

$$S_i^* = P_i - jQ_i = V_i^* I_i = V_i^* \sum_{k=1}^{n} y_{ik} V_k \ \left(i = 1,\ 2, \ldots, n\right) \tag{8.14}$$

From the above set of n complex equations, we obtain the real and imaginary components similar to three-bus case described earlier:

$$P_i = \sum_{k=1}^{n} |y_{ik}||V_i||V_k| \cos\left(\delta_k - \delta_i + \gamma_{ik}\right) \left(i = 1,\ 2, \ldots, n\right)$$

$$Q_i = -\sum_{k=1}^{n} |y_{ik}||V_i||V_k| \sin\left(\delta_k - \delta_i + \gamma_{ik}\right) \left(i = 1,\ 2, \ldots, n\right) \tag{8.15}$$

We can also rewrite the complex form of the S_i* power flow equation (8.14) as follows:

$$P_i - jQ_i = V_i^* y_{ii} V_i + V_i^* \sum_{\substack{k=1 \\ k \neq i}}^{n} y_{ik} V_k \ \left(i = 1,\ 2, \ldots, n\right)$$

$$V_i = \frac{1}{y_{ii}} \left[\frac{P_i - jQ_i}{V_i^*} - \sum_{\substack{k=1 \\ k \neq i}}^{n} y_{ik} V_k \right] \left(i = 1,\ 2, \ldots, n\right) \tag{8.16}$$

The power flow equations in the above form of equation (8.16) are used for the calculation of bus voltages using Gauss–Seidel numerical solution method described later in this chapter.

Example 8.1

The transmission line between buses 1 and 2 in a simplified two-bus system has a line voltage of 345 [kV] and a length of 180 [km]. The line parameters are as follows:

$$R = 10^{-4}\ [\Omega / \mathrm{m}]$$

$$L = 10^{-6}\ [\mathrm{H} / \mathrm{m} / \phi]$$

$$C = 10 \times 10^{-12}\ [\mathrm{F} / \mathrm{m} / \phi]$$

The system base apparent power and base voltage values are as follows:

$$\left|\boldsymbol{S}_{\mathbf{base},3\phi}\right| = 400\ [\mathrm{MVA}]$$

$$\left|\boldsymbol{V}_{\mathbf{base},\,LL}\right| = 345\ [\mathrm{kV}]$$

Find the $\boldsymbol{Y}_{\mathbf{bus}}$ matrix for the simplified two-bus system expressing all admittances in per unit.

Solution

We have:

$$R = 0.1\ [\Omega / \mathrm{km}]$$

$$\omega L = 0.377\ [\Omega / \mathrm{km}]$$

$$\frac{1}{\omega C} = 0.265\ [\mathrm{M}\Omega / \mathrm{km}]$$

$$\left|\boldsymbol{S}_{\mathbf{base},\phi}\right| = 133.33\ [\mathrm{MVA}]$$

$$\left|\boldsymbol{V}_{\mathbf{base},\phi}\right| = 199.19\ [\mathrm{kV}]$$

The base admittance is calculated from the above base values:

$$|Y_{\text{base}}| = \frac{133.33}{199.19^2} = 0.00336\ [\Omega^{-1}/\text{phase}] = 0.00336\ [℧/\text{phase}]$$

We thus have:

$$Y_{s12} = \frac{1}{Z_{s12}} = \frac{1}{180(0.1 + j0.377)}$$

$$= 0.00365 - j0.0138\ [℧/\text{phase}] = 1.08654 - j4.09627\ [\text{pu}\ ℧/\text{phase}]$$

$$Y_{p12} = \frac{1}{-j0.265 \times 10^6} \times 90 = j0.00034\ [℧/\text{phase}]$$

$$= j0.10105\ [\text{pu}\ ℧/\text{phase}]$$

We next calculate Y_{bus} matrix elements:

$$y_{11} = y_{22} = Y_{p12} + Y_{s12} = 1.08654 - j4.09627 + j0.10105$$

$$= 4.1403\angle -74.7858°$$

$$y_{12} = y_{21} = -Y_{s12} = -1.08654 + j4.09627 = 4.2379\angle 104.8560°$$

We then have:

$$Y_{bus} = \begin{bmatrix} 4.1403\angle -74.7858° & 4.2379\angle 104.8560° \\ 4.2379\angle 104.8560° & 4.1403\angle -74.7858° \end{bmatrix} [\text{pu}]$$

Example 8.2

The transmission lines between buses 1 and 2, buses 2 and 3, and buses 3 and 1 in Figure 8.1 are 345 [kV] transmission lines having lengths of 180, 100, and 160 [km], respectively. The transmission line parameters are similar to the previous two-bus example and they are as follows:

$$R = 10^{-4}\ [\Omega/\text{m}]$$

$$L = 10^{-6}\ [\text{H}/\text{m}/\phi]$$

$$C = 10 \times 10^{-12}\ [\text{F}/\text{m}/\phi]$$

The system base apparent power and base voltage values are also similar to the previous two-bus example:

$$|S_{base,3\phi}| = 400\ [\text{MVA}]$$

$$|V_{base,LL}| = 345\ [\text{kV}]$$

Find the Y_{bus} matrix for the three-bus system expressing all admittances in per unit.

Solution

We have following transmission line parameters similar to the previous example:

$$R = 0.1\ [\Omega / \text{km}]$$

$$\omega L = 0.377\ [\Omega / \text{km}]$$

$$\frac{1}{\omega C} = 0.265\ [\text{M}\Omega / \text{km}]$$

We also have:

$$|S_{base,\phi}| = 133.33\ [\text{MVA}]$$

$$|V_{base,\phi}| = 199.19\ [\text{kV}]$$

The base admittance is calculated as earlier from the above base values:

$$|Y_{base}| = \frac{133.33}{199.19^2} = 0.00336\ [\mho / \text{phase}]$$

We thus have:

$$Y_{s12} = \frac{1}{Z_{s12}} = \frac{1}{180(0.1 + j0.377)}$$
$$= 0.00365 - j0.0138\ [\mho / \text{phase}] = 1.087 - j4.096\ [\text{pu}\ \mho / \text{phase}]$$

$$Y_{p12} = \frac{1}{-j0.265 \times 10^6} \times 90 = j0.00034\ [\mho / \text{phase}] = j0.101\ [\text{pu}\ \mho / \text{phase}]$$

$$Y_{s23} = \frac{1}{Z_{s23}} = \frac{1}{100(0.1 + j0.377)}$$
$$= 0.00657 - j0.0248\ [\mho / \text{phase}] = 1.956 - j7.375\ [\text{pu}\ \mho / \text{phase}]$$

$$Y_{p23} = \frac{1}{-j0.265 \times 10^6} \times 50 = j0.000189\ [\mho / \text{phase}] = j0.0562\ [\text{pu}\ \mho / \text{phase}]$$

$$Y_{s13} = \frac{1}{Z_{s13}} = \frac{1}{160(0.1 + j0.377)}$$

$$= 0.00411 - j0.0155\ [\mho/\text{phase}] = 1.223 - j4.61\ [\text{pu}\ \mho/\text{phase}]$$

$$Y_{p13} = \frac{1}{-j0.265 \times 10^6} \times 80 = j0.0003\ [\mho/\text{phase}] = j0.0898\ [\text{pu}\ \mho/\text{phase}]$$

We next calculate Y_{bus} matrix elements:

$$y_{11} = Y_{p12} + Y_{p13} + Y_{s12} + Y_{s13} = 2.31 - j8.515 = 8.82\angle -74.82^\circ$$

$$y_{22} = Y_{p12} + Y_{p23} + Y_{s12} + Y_{s23} = 3.043 - j11.3138 = 11.72\angle -74.95^\circ$$

$$y_{33} = Y_{p13} + Y_{p23} + Y_{s13} + Y_{s23} = 3.179 - j11.839 = 12.26\angle -74.97^\circ$$

$$y_{12} = y_{21} = -Y_{s12} = -1.087 + j4.096 = 4.23\angle 104.86^\circ$$

$$y_{23} = y_{32} = -Y_{s23} = -1.956 + j7.375 = 7.63\angle 104.85^\circ$$

$$y_{31} = y_{13} = -Y_{s13} = -1.223 + j4.61 = 4.77\angle 104.86^\circ$$

We then have:

$$Y_{\text{bus}} = \begin{bmatrix} 8.82\angle -74.82^\circ & 4.23\angle 104.86^\circ & 4.77\angle 104.86^\circ \\ 4.23\angle 104.86^\circ & 11.72\angle -74.95^\circ & 7.63\angle 104.85^\circ \\ 4.77\angle 104.86^\circ & 7.63\angle 104.85^\circ & 12.26\angle -74.97^\circ \end{bmatrix} [\text{pu}]$$

8.3 SPECIFICATION OF VARIABLES FOR THE SOLUTION OF POWER FLOW EQUATIONS

As mentioned earlier, the solution of the three-bus system power flow equations requires specification of 12 out of 18 variables [9,11]. The remaining variables can then be determined from the six power flow equations. The power flow equation variables for the three-bus system are P_{G1}, P_{G2}, P_{G3}, Q_{G1}, Q_{G2}, Q_{G3}, P_{D1}, P_{D2}, P_{D3}, Q_{D1}, Q_{D2}, Q_{D3}, $|V_1|$, $|V_2|$, $|V_3|$, δ_1, δ_2, and δ_3.

The customer demands P_{D1}, P_{D2}, P_{D3}, Q_{D1}, Q_{D2}, and Q_{D3} are assumed to be known. We further choose $\delta_1 = 0$, thereby designating the voltage of bus 1 as our reference phasor. We then have a total of 11 unknowns: P_{G1}, P_{G2}, P_{G3}, Q_{G1}, Q_{G2}, Q_{G3}, $|V_1|$, $|V_2|$, $|V_3|$, δ_2, and δ_3.

It is not possible to initially specify all six generation variables P_{G1}, P_{G2}, P_{G3}, Q_{G1}, Q_{G2}, and Q_{G3} since the system losses P_L and Q_L are not known. We can, however, specify four of the six-generation variables such as P_{G2}, P_{G3}, Q_{G2}, and Q_{G3}, and we then have P_{G1} and Q_{G1} as unknowns.

We have to specify one of the remaining seven unknowns, and then it will be possible to solve the power flow equations for the remaining six variables. If we specify $|V_1|$, we then have following six unknowns: P_{G1}, Q_{G1}, $|V_2|$, $|V_3|$, δ_2, and δ_3.

It is also possible to specify any or both of the other bus voltages $|V_2|$ and $|V_3|$. The variables $|V_2|$ and Q_{G2} are strongly related. When we increase Q_{G2}, $|V_2|$ will rise. Similarly, variables $|V_3|$ and Q_{G3} are strongly related, and an increase in Q_{G2} results in a rise in $|V_3|$. If $|V_2|$ is specified, Q_{G2} is removed from the list of knowns and is added to the list of unknowns. If $|V_3|$ is specified, Q_{G3} is removed from the list of knowns and is added to the list of unknowns. In all cases, the number of unknowns must equal to the number of equations and then it is possible to solve the power flow equations.

From the above description, we find that there can be at least three types of buses depending upon the bus variables specified:

1. Reference or swing bus (P_{Gi}, Q_{Gi} unknown)$\left(|V_i|, \delta_i, P_{Di}, Q_{Di}\text{ known}\right)$
2. Load bus $\left(|V_i|, \delta_i\text{ unknown}\right)\left(P_{Gi}, Q_{Gi}, P_{Di}, Q_{Di}\text{ known}\right)$
3. Voltage control bus $\left(Q_{Gi}, \delta_i\text{ unknown}\right)\left(|V_i|, P_{Gi}, P_{Di}, Q_{Di}\text{ known}\right)$

8.4 ITERATIVE COMPUTATION PROCEDURE OF THE POWER FLOW EQUATIONS

The power flow equations obtained are nonlinear and they are solved using numerical iterative procedure [9,11,38]. In this iterative procedure, we assume initial solution values $V_i^{(0)}$ for all buses whose voltages are unknown. These assumed initial solution values for unknown bus voltages along with known bus voltage values are substituted in power flow equations to obtain more accurate first solution values $V_i^{(1)}$ for unknown bus voltages. These first solution values for unknown bus voltages along with known bus voltage values are next substituted in power flow equations to obtain still more accurate second solution values $V_i^{(2)}$ for unknown bus voltages. This process is repeated to obtain subsequent solution values for unknown bus voltages that are closer to actual voltage values. We stop the calculations when voltage values obtained in any stage are nearly similar to voltage values obtained in the previous stage and do not differ by more than 10^{-3} or 10^{-4} $[\text{pu}]$.

The calculations performed for any bus i will depend on the bus type. The voltage magnitude and phase values of the reference bus are known. The voltage-magnitude values of voltage control buses are known; however, their phase-angle values are unknown. The load buses have voltage-magnitude values and phase-angle values unknown. Prior to starting calculations for unknown bus voltage magnitudes and phase angles using power flow equations, initial solution values are assumed for all unknown bus voltage magnitudes and phase angles. All buses, therefore, have numerical values $V_i^{(0)} = \left|V_i^{(0)}\right| \angle \delta_i^{(0)}$ assigned prior to the start of the calculations.

In first iteration, numerical computations are performed on power flow equations to obtain updated values $V_i^{(1)} = \left|V_i^{(1)}\right| \angle \delta_i^{(1)}$ for bus voltages. Since reference bus has voltage-magnitude and phase-angle values known, calculations are not performed for this bus to obtain updated bus voltage values. For the load buses, P_i and Q_i values are known, and voltage-magnitude and phase-angle values are unknown. Calculations are, therefore, performed for the load buses to obtain updated bus voltage-magnitude and phase-angle values. Voltage control buses have voltage magnitude known and phase-angle values unknown. These buses also have P_i known and Q_i unknown. Q_i

values are, therefore, computed during each iteration for these buses using equation (8.15). The Q_i values obtained are then used in equation (8.16) to obtain updated phase-angle values for the voltage control buses.

The reference bus 1 has voltage-magnitude and phase-angle values known, and P_1 and Q_1 values unknown. After all bus voltage values have been obtained from numerical computations performed on power flow equations as described above, P_1 and Q_1 values are obtained for the reference bus using equation (8.14).

After obtaining all bus voltages, we next determine power flows in all transmission lines.

The transmission line between buses i and j is represented using its equivalent π configuration with series admittance $\boldsymbol{Y}_{sij}$ and two parallel shunt admittances $\boldsymbol{Y}_{pij}$. The transmission line current flowing from bus i to bus j measured at bus i is:

$$\boldsymbol{I}_{ij} = \boldsymbol{I}_{sij} + \boldsymbol{I}_{pij} = \left(\boldsymbol{V}_i - \boldsymbol{V}_j\right)\boldsymbol{Y}_{sij} + \boldsymbol{V}_i\boldsymbol{Y}_{pij} \tag{8.17}$$

The transmission line powers flowing between buses i and j measured at bus i and at bus j are:

$$\begin{aligned} \boldsymbol{S}_{ij} &= P_{ij} + jQ_{ij} = \boldsymbol{V}_i\boldsymbol{I}_{ij}^* = \boldsymbol{V}_i\left(\boldsymbol{V}_i^* - \boldsymbol{V}_j^*\right)\boldsymbol{Y}_{sij}^* + \left|\boldsymbol{V}_i\right|^2 \boldsymbol{Y}_{pij}^* \\ \boldsymbol{S}_{ji} &= P_{ji} + jQ_{ji} = \boldsymbol{V}_j\boldsymbol{I}_{ji}^* = \boldsymbol{V}_j\left(\boldsymbol{V}_j^* - \boldsymbol{V}_i^*\right)\boldsymbol{Y}_{sij}^* + \left|\boldsymbol{V}_j\right|^2 \boldsymbol{Y}_{pij}^* \end{aligned} \tag{8.18}$$

The power loss in the transmission line between buses i and j is simply obtained by summing $\boldsymbol{S}_{ij}$ and $\boldsymbol{S}_{ji}$. The Gauss–Seidel algorithm commonly used to solve the power flow equations is described below.

8.5 GAUSS–SEIDEL ALGORITHM

We first solve a scalar equation of the type shown below using Gauss–Seidel algorithm [11,19,20]:

$$f(x) = 0 \tag{8.19}$$

To do so, we rewrite the above function in the following form:

$$x = F(x) \tag{8.20}$$

To start the iterative solution procedure, we assume initial value $x^{(0)}$ for x. The updated values for x are obtained using the following iterative formula:

$$x^{(v+1)} = F\left(x^{(v)}\right) \tag{8.21}$$

We stop the calculations when the value of x obtained in any stage is nearly similar to the value obtained in the previous stage and does not differ by more than some specified convergence parameter.

The procedure to solve n-dimensional equation using Gauss–Seidel algorithm is similar to that described above for scalar equation. The n-dimensional equation is:

$$\boldsymbol{f}(\boldsymbol{x}) = \boldsymbol{0} \tag{8.22}$$

To obtain solution, we rewrite the above function in the following form:

$$\boldsymbol{x} = \boldsymbol{\mathcal{F}}(\boldsymbol{x}) \tag{8.23}$$

We then have, similar to the scalar iteration equation obtained above, the following iteration formula for the n-dimensional equation:

$$\boldsymbol{x}^{(\mathcal{J}+1)} = \boldsymbol{\mathcal{F}}\left(\boldsymbol{x}^{(\mathcal{J})}\right) \tag{8.24}$$

Example 8.3

Solve the following equation by Gauss–Seidel method:

$$x^2 - 8x + 2 = 0$$

Solution

To solve, we rewrite the above function in the following form:

$$x = \frac{1}{8}x^2 + \frac{1}{4} = F(x)$$

We use the initial estimate $x^{(0)} = 1$. We then obtain in succeeding iterations:

$$\text{First iteration}: x^{(1)} = F(1) = \frac{1}{8} \times 1^2 + \frac{1}{4} = 0.375$$

$$\text{Second iteration}: x^{(2)} = F(0.375) = 0.2676$$

$$\text{Third iteration}: x^{(3)} = F(0.2676) = 0.2590$$

$$\text{Fourth iteration}: x^{(4)} = F(0.2590) = 0.2584$$

$$\text{Fifth iteration}: x^{(4)} = F(0.2584) = 0.2583$$

$$\text{Sixth iteration}: x^{(4)} = F(0.2583) = 0.2583$$

We stop since iterations 5 and 6 yield similar values of x. We therefore have the root $x = 0.258$ correct to three decimal digits.

8.6 GAUSS–SEIDEL METHOD APPLIED TO POWER FLOW EQUATIONS

From the power flow equations for n-bus system described earlier [9,11,38], we have:

$$V_i = \frac{1}{y_{ii}}\left[\frac{P_i - jQ_i}{V_i^*} - \sum_{\substack{k=1\\k\neq i}}^{n} y_{ik}V_k\right]\left(i = 2,\ldots,n\right) \tag{8.16}$$

Since the previous equation has the form $\boldsymbol{x} = \mathcal{F}(\boldsymbol{x})$ described in the previous section, we have our Gauss–Seidel (G-S) algorithm:

$$V_i^{(v+1)} = \frac{1}{y_{ii}}\left[\frac{P_i - jQ_i}{\left(V_i^{(v)}\right)^*} - \sum_{\substack{k=1\\k\neq i}}^{n} y_{ik}V_k^{(v)}\right]\left(i = 2,\ldots,n\right) \tag{8.25}$$

Since bus 1 is assumed to be the reference bus, V_1 is specified. Computations are, therefore, performed to obtain voltage values for buses 2, …, n.

Example 8.4

Figure 8.2 depicts the specified loads at each bus and the voltage specification at bus 1 for a simplified two-bus system. The 345 [kV] transmission line is assumed to have the $\boldsymbol{Y}_{\mathbf{bus}}$ data:

$$\boldsymbol{Y}_{\mathbf{bus}} = \begin{bmatrix} 1.96\angle -75.26^\circ & 2.06\angle 104.04^\circ \\ 2.06\angle 104.04^\circ & 1.96\angle -75.26^\circ \end{bmatrix} [\text{pu}]$$

Bus 1 is a reference bus and bus 2 is a load bus.

Apply the G-S method to the simple two-bus system to find $V_2 = |V_2| \angle \delta_2$. Also, determine P_{G1} and Q_{G1}, and real power loss in the transmission line.

Solution

From the bus loads specified in Figure 8.2, we determine the bus powers:

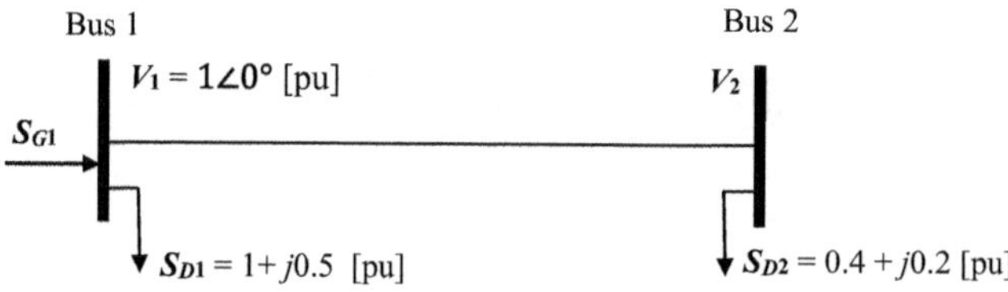

FIGURE 8.2 Example power system.

$$S_1 = (P_{G1} - 1) + j(Q_{G1} - 0.5)\ [\text{pu}]$$

$$S_2 = -0.4 - j0.2\ [\text{pu}]$$

In this case, $n = 2$ and $V_i^{(v+1)}$ equation obtained earlier reduces to one complex equation:

$$V_2^{(v+1)} = \frac{1}{y_{22}}\left[\frac{P_2 - jQ_2}{\left(V_2^{(v)}\right)^*} - y_{21}V_1\right]$$

We have $V_1 = 1\angle 0°\ [\text{pu kV}]$.

The G-S algorithm after substituting numerical values is:

$$V_2^{(v+1)} = \left(\frac{1}{1.96\angle -75.26°}\right)\left[\frac{-0.4 + j0.2}{\left(V_2^{(v)}\right)^*} - 2.06\angle 104.04° \times 1\angle 0°\right]$$

$$= (0.510\angle 75.26°)\left[\frac{0.447\angle 153.44°}{\left(V_2^{(v)}\right)^*} - 2.06\angle 104.04°\right]$$

We assume:

$$V_2^{(0)} = 0.95\angle -10°\ [\text{pu kV}]$$

Our algorithm then yields:

$$V_2^{(1)} = (0.510\angle 75.26°)\left[\frac{0.447\angle 153.44°}{0.95\angle 10°} - 2.06\angle 104.04°\right]$$

$$= 0.879\angle -10.69°\ [\text{pu kV}]$$

The next iteration yields:

$$V_2^{(2)} = (0.510\angle 75.26°)\left[\frac{0.447\angle 153.44°}{0.879\angle -10.69°} - 2.06\angle 104.04°\right]$$

$$= 0.864\angle -11.53°\ [\text{pu kV}]$$

From further iterations, we have:

$$V_2^{(3)} = 0.858\angle -11.59°\ [\text{pu kV}]$$

$$V_2^{(4)} = 0.857\angle -11.67°\ [\text{pu kV}]$$

$$V_2^{(5)} = 0.856\angle -11.68° \ [\text{pu kV}]$$

$$V_2^{(6)} = 0.856\angle -11.68° \ [\text{pu kV}]$$

The voltage convergence thus required six iterations. We thus have:

$$V_2 = 0.856\angle -11.68° \ [\text{pu kV}]$$

To determine $\boldsymbol{S}_1$ and $\boldsymbol{S}_2$, we utilize the following equations described earlier:

$$S_1^* = P_1 - jQ_1 = V_1^* \sum_{k=1}^{2} y_{1k} V_k$$

$$S_2^* = P_2 - jQ_2 = V_2^* \sum_{k=1}^{2} y_{2k} V_k$$

$$S_1 = P_1 + jQ_1 = 0.426 + j0.134 \ [\text{pu MVA}]$$

$$S_2 = P_2 + jQ_2 = -0.4 - j0.2 \ [\text{pu MVA}]$$

We have from equation (8.18):

$$S_{12} = P_{12} + jQ_{12} = 0.426 + j0.134 \ [\text{pu MVA}]$$

$$S_{21} = P_{21} + jQ_{21} = -0.4 - j0.2 \ [\text{pu MVA}]$$

We further have:

$$P_{G1} = P_1 + P_{D1} = 0.426 + 1 = 1.426 \ [\text{pu MW}]$$

$$Q_{G1} = Q_1 + Q_{D1} = 0.134 + 0.5 = 0.634 \ [\text{pu VAR}]$$

Total real power loss:

$$P_{Loss} = P_{12} + P_{21} = 0.026 \ [\text{pu MW}]$$

Example 8.5

In the three-bus system of Figure 8.1 and Example 8.2, the specified loads at buses 1, 2, and 3, and the voltage specification at bus 1 are as follows:

$$S_{D1} = 1 + j0.5 \ [\text{pu MVA}]$$

$$S_{D2} = 0.4 + j0.2 \ [\text{pu MVA}]$$

$$S_{D3} = 0.8 + j0.6 \ [\text{pu MVA}]$$

$$V_1 = 1\angle 0° \ [\text{pu kV}]$$

Bus 1 is a reference bus receiving power S_{G1} from the generator connected to it. Buses 2 and 3 are load buses and not connected to generators ($S_{G2} = S_{G3} = 0$).

The 345 [kV] transmission system is assumed to have the following Y_{bus} data obtained earlier:

$$Y_{\text{bus}} = \begin{bmatrix} 8.82\angle -74.82° & 4.23\angle 104.86° & 4.77\angle 104.86° \\ 4.23\angle 104.86° & 11.72\angle -74.95° & 7.63\angle 104.85° \\ 4.77\angle 104.86° & 7.63\angle 104.85° & 12.26\angle -74.97° \end{bmatrix} [\text{pu}]$$

Apply the G-S method to the simple three-bus power system to find $V_2 = |V_2|\angle\delta_2$ and $V_3 = |V_3|\angle\delta_3$.

Solution

From the bus loads specified, we find the bus powers:

$$S_1 = (P_{G1} - 1) + j(Q_{G1} - 0.5) \ [\text{pu}]$$

$$S_2 = -0.4 - j0.2 \ [\text{pu}]$$

$$S_3 = -0.8 - j0.6 \ [\text{pu}]$$

In this case, $n = 3$ and $V_i^{(\nu+1)}$ equation obtained above reduces to two complex equations:

$$V_2^{(\nu+1)} = \frac{1}{y_{22}}\left[\frac{P_2 - jQ_2}{\left(V_2^{(\nu)}\right)^*} - y_{21}V_1 - y_{23}V_3^{(\nu)}\right]$$

$$V_3^{(\nu+1)} = \frac{1}{y_{33}}\left[\frac{P_3 - jQ_3}{\left(V_3^{(\nu)}\right)^*} - y_{31}V_1 - y_{32}V_2^{(\nu)}\right]$$

We have:

$$\boldsymbol{V}_1 = 1\angle 0^\circ \;[\text{pu kV}]$$

We further assume:

$$\boldsymbol{V}_2^{(0)} = \boldsymbol{V}_3^{(0)} = 0.95\angle -10^\circ \;[\text{pu kV}]$$

The G-S algorithm after substituting numerical values becomes:

$$\boldsymbol{V}_2^{(\nu+1)}$$

$$= \left(\frac{1}{11.72\angle -74.95^\circ}\right)\left[\frac{-0.4 + j0.2}{\left(\boldsymbol{V}_2^{(\nu)}\right)^*} - 4.23\angle 104.86^\circ \times 1\angle 0^\circ - 7.63\angle 104.85^\circ \times \boldsymbol{V}_3^{(\nu)}\right]$$

$$= (0.0854\angle 74.95^\circ)\left[\frac{0.447\angle 153.44^\circ}{\left(\boldsymbol{V}_2^{(\nu)}\right)^*} - 4.23\angle 104.86^\circ - 7.63\angle 104.85^\circ \times \boldsymbol{V}_3^{(\nu)}\right]$$

$$\boldsymbol{V}_3^{(\nu+1)}$$

$$= \left(\frac{1}{12.26\angle -74.97^\circ}\right)\left[\frac{-0.8 + j0.6}{\left(\boldsymbol{V}_3^{(\nu)}\right)^*} - 4.77\angle 104.86^\circ \times 1\angle 0^\circ - 7.63\angle 104.85^\circ \times \boldsymbol{V}_2^{(\nu)}\right]$$

$$= (0.0816\angle 74.97^\circ)\left[\frac{1.0\angle 143.13^\circ}{\left(\boldsymbol{V}_3^{(\nu)}\right)^*} - 4.77\angle 104.86^\circ - 7.63\angle 104.85^\circ \times \boldsymbol{V}_2^{(\nu)}\right]$$

We then obtain:

$$\boldsymbol{V}_2^{(1)}$$

$$= (0.0854\angle 74.95^\circ)\left[\frac{0.447\angle 153.44^\circ}{0.95\angle 10^\circ} - 4.23\angle 104.86^\circ - 7.63\angle 104.85^\circ \times 0.95\angle -10^\circ\right]$$

$$= 0.949\angle -8.23^\circ \;[\text{pu kV}]$$

$$\boldsymbol{V}_3^{(1)}$$

$$= (0.0816\angle 74.97^\circ)\left[\frac{1.0\angle 143.13^\circ}{0.95\angle 10^{\circ *}} - 4.77\angle 104.86^\circ - 7.63\angle 104.85^\circ \times 0.95\angle -10^\circ\right]$$

$$= 0.907\angle -9.27^\circ \;[\text{pu kV}]$$

From the next two iterations, we obtain:

$$V_2^{(2)} = 0.922\angle -7.75° \left[\text{pu kV}\right]$$

$$V_3^{(2)} = 0.904\angle -8.33° \left[\text{pu kV}\right]$$

$$V_2^{(3)} = 0.920\angle -7.22° \left[\text{pu kV}\right]$$

$$V_3^{(3)} = 0.888\angle -8.10° \left[\text{pu kV}\right]$$

The voltage convergence requires 13 iterations. We finally have:

$$V_2 = 0.899\angle -6.88° \left[\text{pu kV}\right]$$

$$V_3 = 0.872\angle -7.73° \left[\text{pu kV}\right]$$

Example 8.6

In two-bus system of Example 8.4, the capacitor is connected at bus 2 as shown in Figure 8.3.

The bus loads and reference bus voltage for the two-bus system are same as in Example 8.4. It is, however, required to maintain bus voltage magnitude $|V_2|$ equal to $|V_2| = 0.95\left[\text{pu kV}\right]$ by supplying reactive power.

Determine δ_2 and thus the voltage phasor V_2.

Solution

We utilize our two-bus algorithm as earlier in the present example:

$$V_2^{(v+1)} = \frac{1}{y_{22}}\left[\frac{P_2 - jQ_2}{\left(V_2^{(v)}\right)^*} - y_{21}V_1\right]$$

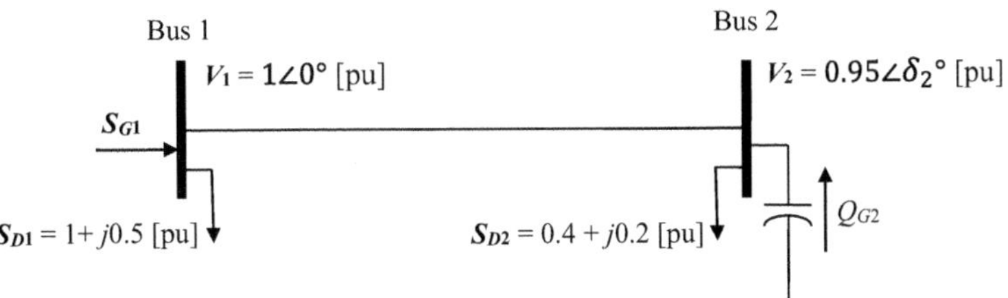

FIGURE 8.3 Example power system.

The above algorithm for V_2, however, must be corrected in two ways.

The first correction is that the bus power Q_2 is computed from the equation described earlier and shown below:

$$Q_2 = -\sum_{k=1}^{2} |y_{2k}||V_2||V_k| \sin(\delta_k - \delta_2 + \gamma_{2k})$$

The second correction is that V_2^* magnitude is maintained at the specified value of $|V_2| = 0.95\ [\text{pu kV}]$.

Our algorithm therefore becomes:

$$V_2^{(v+1)} = (0.510\angle 75.26°)\left[\frac{-0.40 - jQ_2^{(v)}}{0.95\angle -\delta_2^{(v)}} - 2.06\angle 104.04°\right]$$

For $Q_2^{(v)}$, we have:

$$Q_2^{(v)} = -\sum_{k=1}^{2} |y_{2k}|\left|V_2^{(v)}\right||V_k| \sin\left(\delta_k - \delta_2^{(v)} + \gamma_{2k}\right)$$

$$Q_2^{(v)} = 1.711 - 1.957 \sin\left(104.04° - \delta_2^{(v)}\right)$$

We assume:

$$V_2^{(0)} = 0.95\angle -10°\ [\text{pu kV}]$$

From $Q_2^{(v)}$ equation above, we compute:

$$Q_2^{(0)} = 1.711 - 1.957 \sin(104.04° + 10°)$$
$$= -0.0766\ [\text{pu MVAR}]$$

After substituting $Q_2^{(0)}$ into $V_2^{(v+1)}$ equation, we obtain:

$$V_2^{(1)} = (0.510\angle 75.26°)\left[\frac{-0.40 + j0.0766}{0.95\angle 10°} - 2.06\angle 104.04°\right]$$
$$= 0.943\angle -11.67°\ \text{pu kV}$$

From $Q_2^{(v)}$ equation, we obtain:

$$Q_2^{(1)} = 1.711 - 1.957 \sin(104.04° + 11.67°)$$
$$= -0.0526\ [\text{pu MVAR}]$$

More iterations yield:

$$V_2^{(2)} = 0.950\angle -11.70^\circ \ [\text{pu kV}]$$

$$Q_2^{(2)} = -0.0521 \ [\text{pu MVAR}]$$

$$V_2^{(3)} = 0.950\angle -11.71^\circ \ [\text{pu kV}]$$

$$Q_2^{(3)} = -0.0520 \ [\text{pu MVAR}]$$

$$V_2^{(4)} = 0.950\angle -11.71^\circ \ [\text{pu kV}]$$

$$Q_2^{(4)} = -0.0520 \ [\text{pu MVAR}]$$

The voltage convergence thus required four iterations.

To summarize, the bus voltages are:

$$V_1 = 1\angle 0^\circ \ [\text{pu kV}]$$

$$V_2 = 0.950\angle -11.71^\circ \ [\text{pu kV}]$$

8.7 POWER FLOW CONTROL

Being able to compute the power flows in any power system network is simply not enough. It must be possible to control the power flows in the network too. The methods used for power flow control in power systems are prime mover torque and excitation control of the generators, connecting and disconnecting series and shunt capacitors and reactors, tap changing of regular transformers, and voltage-magnitude and phase-angle regulating transformers [9,11,16].

Only few buses in power system network have generators connected to them. The real power output of a generator is proportional to the prime mover or turbine torque, which is regulated for the control of real power flow in power system network. The emf produced in a generator is similarly proportional to the rotor field current, which is thus regulated for voltage control and indirectly for reactive power flow control. Capacitors and reactors are also connected and disconnected as required for controlling the reactive power flows in power system network and thus the voltage profile as described earlier.

Tap-changing and voltage-magnitude-regulating transformers are used for the control of bus voltages, as well as reactive power flows on transmission lines to which they are connected. Similarly, phase-angle regulating transformers are used to control bus phase angles as well as real power flows on transmission lines to which they are connected. Adjustments in transformer settings, however, result in the alteration of $\boldsymbol{Y}_{\text{bus}}$ matrix unlike adjustments in generation settings which have no effect on the $\boldsymbol{Y}_{\text{bus}}$ matrix.

One must choose generation levels and other power flow control settings described above properly using engineering judgment to meet the specified requirements about power system equipment loading and network voltage profile in the best possible way.

The reader is requested to refer to the standard power system texts mentioned in the list of references to learn further about power flow analysis and power flow control methods in power systems.

8.8 ECONOMIC OPERATION OF POWER SYSTEMS

The power requirements of all connected loads in a large plant or a power system are met by a number of generators working together. In order to accommodate the increasing demand of the customers, these generators are typically put into service over a number of years as the plant or power system expands. Due to this reason, the fuel energy conversion characteristics of many generators can vary greatly, which implies that certain generators are capable of providing more electrical output per unit of fuel energy input than other generators. In these conditions, distributing the total load requirement among the generators in a way that minimizes total fuel costs is the proper policy. This policy is known as optimal economic dispatch [9].

There are two cases analyzed in optimal economic dispatch. The first is known as the lossless case, in which transmission line losses are not considered in determining the generator load assignments. A large plant whose electrical power requirements are met by numerous locally installed generators is a typical example of this situation. Implementation of the rule of optimal economic dispatch ensures that these generators operate to meet the load demands in a way that minimizes total fuel costs during various load conditions.

The second case of optimal economic dispatch involves the entire power system and all connected generating units. In this case, transmission line losses are significant and they are considered [9]. The load assignments of various generators in the power system in this case can be greatly influenced by transmission line losses. For instance, we assume that a distant load receives power from two power plants over separate transmission lines. One of the generators has a lower power generation cost than the other; however, it has much higher transmission line losses. Implementation of the rule of optimal economic dispatch will direct the second generator to supply most of the power to the load in order to minimize total fuel costs. The first generator has a penalty factor attached to it to account for higher transmission line losses, which affects the load assignment of the generators.

The student reader is requested to refer to the standard power system references mentioned in the list of references for detailed description of economic operation of power systems.

8.9 SUMMARY

The purpose of power flow analysis is the determination of the bus voltages of a power system under balanced three-phase steady-state conditions. The bus voltage values obtained are then used for the determination of transmission line currents, real and reactive power flows, and power system network losses for particular load conditions in the power system. Numerical methods are used in power flow analysis to determine the bus voltages of the power system.

The simplest iterative method used for power flow analysis is the Gauss–Seidel iterative method. In this method, the voltage at each bus is determined from the knowledge of powers supplied to the buses, the power drawn from the buses, and an estimate of all of the other bus voltages in the system. After the determination of all bus voltages, the power flows in various transmission lines of the network and power losses in them are easily computed.

The Gauss–Seidel method is extremely simple; however, it converges slowly for large power systems. Another iterative method commonly utilized is the Newton–Raphson method. It converges rapidly for both large and small power systems in a few iterations. It is not described here due to its relative complexity in comparison to Gauss–Seidel method and to keep the description of power flow analysis simple. The reader is requested to refer to the standard electrical power system texts to learn about it.

The purpose of optimal economic dispatch is to determine the power output of all generators in a large plant or a power system to meet the power requirements of all connected loads in a way that minimizes total fuel costs. There are two cases of optimal economic dispatch. The first is the lossless case in which transmission line losses are not considered. In the second case of optimal economic dispatch, transmission line losses are considered.

REVIEW QUESTIONS

8.1 What is the purpose of power flow analysis?
8.2 Derive the power flow equations for the three-bus power system used in power flow analysis.
8.3 Extend the power flow equations for the three-bus power system to the general n-bus power system.
8.4 What are the different types into which buses are classified in power flow analysis?
8.5 What variables are specified for each type of bus used in power flow analysis?
8.6 Describe the Gauss–Seidel iterative method applied for solving the power flow equations.
8.7 Describe the procedure for the determination of transmission line losses in the Gauss–Seidel iterative method.
8.8 What are the methods used for control of power flow in a power system?
8.9 What is optimal economic dispatch?
8.10 What are the two cases of optimal economic dispatch?

9 Power System Fault Analysis and Protection

9.1 INTRODUCTION

An electric power system operates as a balanced three-phase AC system when it is working properly. A fault occurs in a power system due to any failure that affects the proper flow of current to the loads. Faults result due to the connection and formation of current path between two or more phase conductors, or between one or more phase conductors and ground. The connection which creates a fault can be from physical contact or due to arcing occurring from current flow through a gaseous medium. Currents in the power system resulting from a fault can be extremely high.

The types of faults which occur in an electric power system can be divided into two main categories: symmetrical faults in which the magnitudes of the AC currents in the three phases are equal, and unsymmetrical faults in which the magnitudes of the AC currents in the three phases differ. Symmetrical three-phase fault is an example of symmetrical power system fault. Single line-to-ground fault, line-to-line fault, and double line-to-ground fault are examples of unsymmetrical power system faults.

It is necessary to determine the values of power system voltages and currents during faults for obtaining the settings of protective devices. The purpose of power system protection is to detect and isolate a faulted portion from the rest of the power system so that the unfaulted portion can function properly without damage due to fault current.

This chapter is divided into three main sections: symmetrical fault analysis, unsymmetrical fault analysis, and power system protection.

Part I: Symmetrical Fault Analysis

9.2 SYMMETRICAL FAULTS ON THREE-PHASE SYSTEMS

Balanced three-phase faults and balanced three-phase to-ground faults are examples of symmetrical faults. A balanced three-phase fault is identical to a balanced three-phase to-ground fault as the faulted power system is balanced, and since fault currents in the three phases add to zero, the current in the ground conductor is zero. Balanced three-phase faults occur when there is a simultaneous short circuit across all three phases. Balanced three-phase to-ground faults occur when all three phases are simultaneously shorted together and are grounded. These faults occur relatively infrequently than unsymmetrical faults such as single line-to-ground faults. Symmetrical three-phase faults, however, are the most severe type of faults occurring on power systems. Since the power system after the occurrence of symmetrical

DOI: 10.1201/9781003432340-9

faults is balanced, it can be solved on a per-phase basis. The other two phases will have identical voltages and currents except for the 120° phase shift.

9.2.1 Symmetrical Short Circuit on Synchronous Machine

Prior to proceeding with the calculation of fault current, we first learn about the behavior of the synchronous generator when it is short circuited [VI,33,38]. We start with the understanding of transient behavior of series *R–L* circuit when an AC voltage source is connected across it, which approximately resembles a three-phase short-circuit condition at the terminals of an unloaded synchronous generator. The current that flows in a series *R–L* circuit immediately after connecting an AC voltage source across it has two components: a steady-state sinusoidally varying current component of constant magnitude, and a non-periodic and exponentially decaying current component with time constant *L*/*R* also referred to as DC component of current. The initial value of DC component of the current depends upon the magnitude of AC voltage at the instant of closing of the circuit.

The events which occur after the occurrence of a three-phase short circuit across the terminals of a synchronous generator are similar to and more complex than those described above for series *R–L* circuit. We draw current in one of the phases after the occurrence of three-phase short circuit at the terminals of an unloaded synchronous generator to analyze the effect. The three-phase voltages generated in the synchronous generator are displaced 120° from each other. When the fault occurs, the voltage values in each of the three phases are different. Due to this reason, each of the three phases of the synchronous machine has a different transient DC component of current. Figure 9.1 shows the plot of short-circuit current with respect to time after ignoring DC transient component of current in each phase.

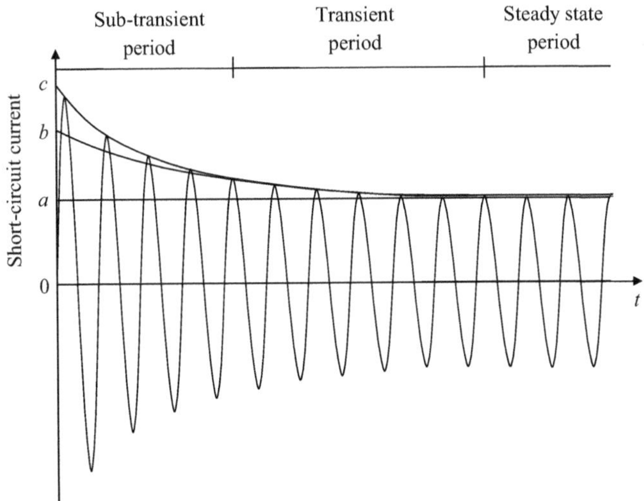

FIGURE 9.1 Short-circuit current of the synchronous generator [38].

In a synchronous generator, there is a reduction in air gap field flux due to flux from the flow of currents in the stator armature windings. This effect is known as armature reaction, and it is described in a previous chapter. Prior to the occurrence of the three-phase short circuit, the armature reaction effect is small due to low values of armature current, and consequently air gap flux has high value. After the occurrence of three-phase short circuit, there is an immediate increase in currents in stator armature windings. The air gap flux, however, does not change instantaneously since eddy currents flowing in the rotor and damper circuits oppose such an occurrence. Consequently, the armature reaction effect is very less initially and reactance due to it is very less too. The initial synchronous reactance therefore has a very small value, and it is nearly equal to armature leakage reactance. The result of this is very high initial current as shown in Figure 9.1. This initial period is called sub-transient period.

The eddy currents flowing in the damper circuit and subsequently in the field circuits decay somewhat after a few cycles. Consequently, partial armature reaction effect results in a reduction in air gap flux and a decrease in stator short-circuit current as shown in Figure 9.1. This period of synchronous generator operation is referred to as transient period.

The eddy currents flowing in the damper circuit decay completely after a few more cycles. The armature reaction effect then results in a further reduction in air gap flux and a decrease in stator short-circuit current. We then have a steady-state period reached as shown in Figure 9.1.

The sub-transient current $|\boldsymbol{I}''|$ is the rms value of current obtained from intercept of current envelope with zero time as shown in Figure 9.1. The sub-transient reactance X_s'' of the synchronous generator therefore is $|\boldsymbol{E}|/|\boldsymbol{I}''|$, where $|\boldsymbol{E}|$ is the generated rms phase voltage of the unloaded synchronous generator.

The transient current $|\boldsymbol{I}'|$ is the rms value of current obtained from intercept of current envelope a few cycles after the occurrence of three-phase fault with zero time. The transient reactance X_s' of the synchronous generator therefore is $|\boldsymbol{E}|/|\boldsymbol{I}'|$.

The steady-state current $|\boldsymbol{I}_{ss}|$ is the rms value of the steady-state short-circuit current. The steady-state reactance X_{ss} of the synchronous generator therefore is $|\boldsymbol{E}|/|\boldsymbol{I}_{ss}|$.

We have the following relationships between sub-transient, transient, and steady-state currents and impedances.

$$|\boldsymbol{I}''| = \frac{0c}{\sqrt{2}} = \frac{|\boldsymbol{E}|}{X_s''} \tag{9.1}$$

$$|\boldsymbol{I}'| = \frac{0b}{\sqrt{2}} = \frac{|\boldsymbol{E}|}{X_s'} \tag{9.2}$$

$$|\boldsymbol{I}_{ss}| = \frac{0a}{\sqrt{2}} = \frac{|\boldsymbol{E}|}{X_{ss}} \tag{9.3}$$

The magnitude of sub-transient current $|\boldsymbol{I}''|$ is much higher than that of steady-state current $|\boldsymbol{I}_{ss}|$. It is due to the fact that reduction in air gap flux due to increase in stator armature current does not occur immediately as described earlier. Large air gap flux therefore induces large voltages in armature windings immediately after the fault. The voltages existing in armature windings are less after steady-state condition is reached. Different voltages induced in synchronous generator during sub-transient, transient, and steady-state periods are accounted for by including different reactances in series with generator no-load voltage $\boldsymbol{E}$ in calculation of sub-transient, transient, and steady-state currents [VI,33,38].

From equations (9.1)–(9.3) described above, it is possible to calculate sub-transient, transient, and steady-state currents if corresponding reactance values are known. For higher accuracy, the synchronous generator resistance is also considered. For three-phase short circuit occurring on an unloaded synchronous generator, the generator is represented by generated no-load voltage $\boldsymbol{E}$ in series with the proper value of synchronous machine reactance as described above. If the power system has external impedance due to other power system components between synchronous generator terminals and fault location, the circuit for fault current calculation must include the external impedance. The procedure for the calculation of sub-transient, transient, and steady-state currents in power systems during a three-phase fault is described using an example below.

Example 9.1

A synchronous generator is connected to the low voltage side of a three-phase Δ–Y transformer as shown in Figure 9.2. The generator is rated 150 [MVA], 24 [kV]. The generator has sub-transient reactance, transient reactance, and steady-state reactance of 0.20, 0.35, and 1.0 [pu], respectively. The transformer is rated 150 [MVA], 24 [kV] Δ/115 [kV] Y with a reactance of 0.08 [pu]. All machine reactances are obtained using machine ratings as base values. Before the fault occurs, the transformer is unloaded and the voltage on the high-voltage side of the transformer is 110 [kV].

a. Find the current supplied by generator before the occurrence of fault.
b. A three-phase short circuit occurs at F_1. Determine the sub-transient short circuit current in generator.
c. A three-phase short circuit occurs at F_2. Determine the transient and steady-state short-circuit current in generator.

Select a base of 150 [MVA] and 115 [kV] on the high-voltage side of the transformer.

Solution

$$\text{Base voltage on the low voltage side} = 24\ [\text{kV}]$$

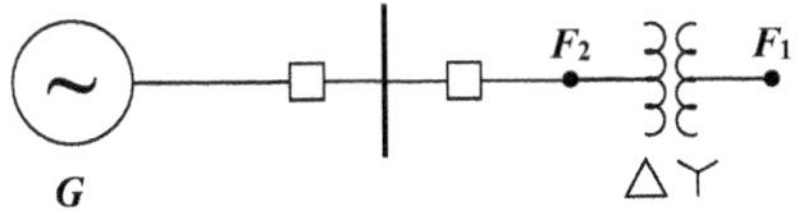

FIGURE 9.2 Example one-line diagram of the faulted power system.

$$\text{Pre-fault voltage on the low voltage side} = \frac{24}{115} \times 110 = 22.96\ [\text{kV}]$$

$$\text{Base current on the low voltage side} = \frac{150 \times 10^3}{\sqrt{3} \times 24} = 3608.44\ [\text{A}]$$

On the selected base for the power system, we have:

Generator:

$$X_s'' = 0.2\ [\text{pu}]$$

$$X_s' = 0.35\ [\text{pu}]$$

$$X_{ss} = 1\ [\text{pu}]$$

$$|\boldsymbol{E}| = \frac{22.96}{24} = 0.96\ [\text{pu}]$$

Transformer:

$$X_T = 0.08\ [\text{pu}]$$

a. The transformer is unloaded.
   We therefore have

$$\text{Current supplied by generator before fault } \boldsymbol{I} = 0$$

b. Fault occurs at F_1.
   Per-unit sub-transient reactance diagram for the power system is shown in Figure 9.3.

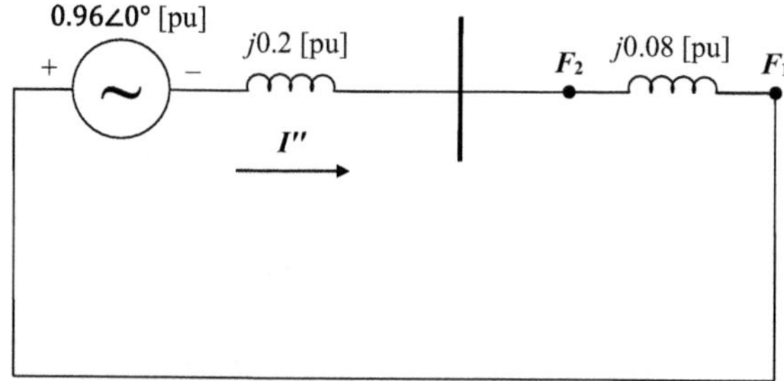

FIGURE 9.3 Per-unit sub-transient reactance diagram.

$$\text{Subtransient current } \boldsymbol{I''} = \frac{0.96\angle 0^\circ}{j0.2 + j0.08} = -j3.42\ [\text{pu}]$$

$$\text{Subtransient current in generator } |\boldsymbol{I''}| = 12333.65\ [\text{A}]$$

c. Fault occurs at F_2.

The per-unit transient reactance diagram for the power system is shown in Figure 9.4.

$$\text{Transient current } \boldsymbol{I'} = \frac{0.96\angle 0^\circ}{j0.35} = -j2.73\ [\text{pu}]$$

$$\text{Transient current in generator } |\boldsymbol{I'}| = 9861.87\ [\text{A}]$$

The per-unit steady-state reactance diagram for the power system is shown in Figure 9.5.

$$\text{Steady state current } \boldsymbol{I_{ss}} = \frac{0.96\angle 0^\circ}{j1.0} = -j0.96\ [\text{pu}]$$

$$\text{Steady state short circuit current in generator } |\boldsymbol{I_{ss}}| = 3453.28\ [\text{A}]$$

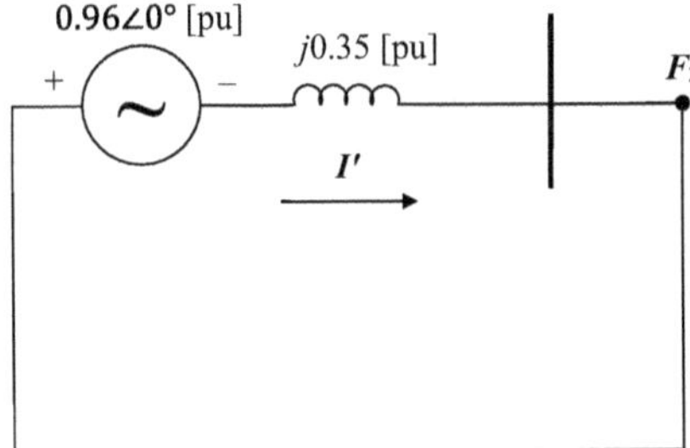

FIGURE 9.4 Per-unit transient reactance diagram.

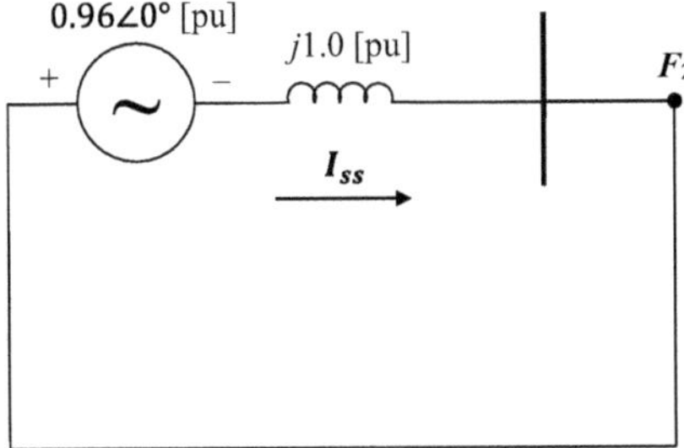

FIGURE 9.5 Per-unit steady-state reactance diagram.

Part II: Unsymmetrical Fault Analysis

9.3 INTRODUCTION TO UNSYMMETRICAL FAULTS

Symmetrical faults on power systems are the most severe; however, the analysis of such faults is simple. Since symmetrical faults are balanced, it is required to consider only one phase for the calculation of power system voltages and currents. The voltages and currents in the other two phases are identical with a phase shift of 120°.

Most of the faults that occur on the power system are unsymmetrical such as single line-to-ground fault, line-to-line fault, and double line-to-ground fault. Occurrence of an unsymmetrical fault results in unsymmetrical currents and voltages in a power system. It means currents and voltages in the three phases of the power system have different magnitudes and unequal phase displacements.

For the analysis of unsymmetrical faults, the method of symmetrical components is utilized for the estimation of power system voltages and currents at various locations as described below.

9.4 SYMMETRICAL COMPONENT METHOD

For the analysis of unsymmetrical faults or unbalanced system conditions, the method of symmetrical components is utilized. The method of symmetrical components converts a three-phase unbalanced circuit to three balanced circuits, so that the system can be represented by three single-phase equivalent sequence networks. The sequence networks thus obtained are then connected in a manner that represents the fault or unbalance condition of the actual power system, and permits calculations of power system voltages and currents for this condition.

9.4.1 SEQUENCE VOLTAGES AND CURRENTS

Using the method of symmetrical components, any set of three unbalanced voltage or current phasors can be resolved into three sets of balanced phasors, called the sequence components. The three phasors of each balanced set have equal magnitude and are displaced by 0°, 120°, or –120°. Figure 9.6 shows the application of this method to the unbalanced voltage phasors of a three-phase circuit.

The components in Figure 9.6 with subscript 1 form the set of positive-sequence components. Similarly, the components with subscripts 0 and 2 form the sets of negative- and zero-sequence components, respectively. The three sets add up to form the original three-phase unbalanced voltage phasors as shown in Figure 9.7.

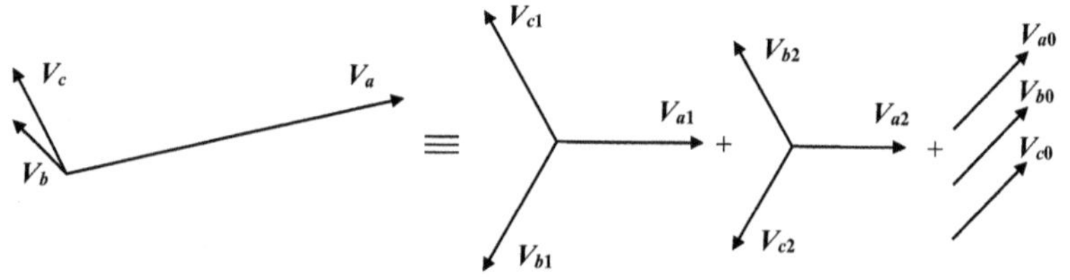

V_c V_a V_b $\equiv$ V_{c1} V_{a1} V_{b1} + V_{b2} V_{a2} V_{c2} + V_{a0} V_{b0} V_{c0}

FIGURE 9.6 Symmetrical components of a three-phase system.

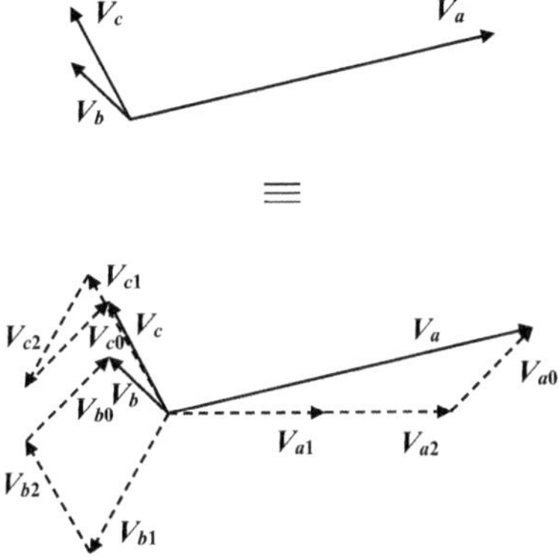

FIGURE 9.7 Graphical summation of the symmetrical components in order to obtain the three-phase unbalanced voltage phasors.

We thus have the following relationships for the voltage phasors of a three-phase system:

$$\begin{aligned} \boldsymbol{V}_a &= \boldsymbol{V}_{a0} + \boldsymbol{V}_{a1} + \boldsymbol{V}_{a2} \\ \boldsymbol{V}_b &= \boldsymbol{V}_{b0} + \boldsymbol{V}_{b1} + \boldsymbol{V}_{b2} \\ \boldsymbol{V}_c &= \boldsymbol{V}_{c0} + \boldsymbol{V}_{c1} + \boldsymbol{V}_{c2} \end{aligned} \tag{9.4}$$

A perfectly balanced system is composed completely of positive-sequence components, which means equal magnitudes, phase sequence *abc* with 120° phase shift between phases. Negative sequence components in a power system indicate the level of system unbalance. This unbalance can be harmful to power system components. Zero-sequence components are related to faults or unbalances involving neutral or ground [3,17,38]. For example, the level of zero-sequence current is directly related to the level of neutral or ground current.

For angle displacement of sequence components, the $\boldsymbol{a}$-operator is commonly utilized which results in counterclockwise rotation of a phasor by 120°, such that:

$$\begin{aligned} \boldsymbol{a} &= 1\angle 120° = 1 \times e^{j120°} = -0.5 + j0.866 \\ \boldsymbol{a}^2 &= 1\angle 240° = 1 \times e^{j240°} = -0.5 - j0.866 \\ \boldsymbol{a}^3 &= 1\angle 360° = 1 \times e^{j360°} = 1\angle 0° = 1 + j0 \end{aligned} \tag{9.5}$$

$$1 + \boldsymbol{a} + \boldsymbol{a}^2 = 0 \tag{9.6}$$

A graphical representation of the properties of the $\boldsymbol{a}$-operator is shown in Figure 9.8.

We can, therefore, write the components of a given sequence in terms of any particular component. We thus have

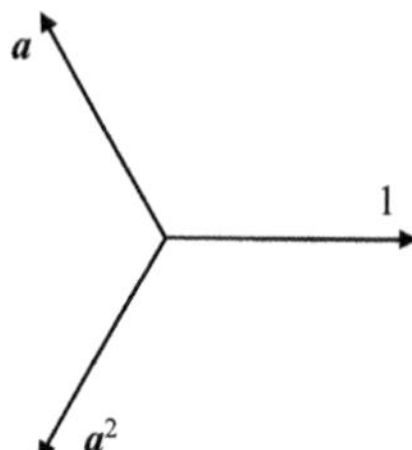

FIGURE 9.8 The a-operator properties.

$$\begin{aligned}
\boldsymbol{V}_{b1} &= \boldsymbol{a}^2\boldsymbol{V}_{a1} \\
\boldsymbol{V}_{c1} &= \boldsymbol{a}\boldsymbol{V}_{a1} \\
\boldsymbol{V}_{b2} &= \boldsymbol{a}\boldsymbol{V}_{a2} \\
\boldsymbol{V}_{c2} &= \boldsymbol{a}^2\boldsymbol{V}_{a2} \\
\boldsymbol{V}_{a0} &= \boldsymbol{V}_{b0} = \boldsymbol{V}_{c0}
\end{aligned} \tag{9.7}$$

In terms of components of phase a, we have

$$\begin{aligned}
\boldsymbol{V}_a &= \boldsymbol{V}_{a0} + \boldsymbol{V}_{a1} + \boldsymbol{V}_{a2} \\
\boldsymbol{V}_b &= \boldsymbol{V}_{a0} + \boldsymbol{a}^2\boldsymbol{V}_{a1} + \boldsymbol{a}\boldsymbol{V}_{a2} \\
\boldsymbol{V}_c &= \boldsymbol{V}_{a0} + \boldsymbol{a}\boldsymbol{V}_{a1} + \boldsymbol{a}^2\boldsymbol{V}_{a2}
\end{aligned} \tag{9.8}$$

We can also write the above relationships in matrix form as below:

$$\begin{bmatrix} \boldsymbol{V}_a \\ \boldsymbol{V}_b \\ \boldsymbol{V}_c \end{bmatrix} = \begin{bmatrix} 1 & 1 & 1 \\ 1 & \boldsymbol{a}^2 & \boldsymbol{a} \\ 1 & \boldsymbol{a} & \boldsymbol{a}^2 \end{bmatrix} \times \begin{bmatrix} \boldsymbol{V}_{a0} \\ \boldsymbol{V}_{a1} \\ \boldsymbol{V}_{a2} \end{bmatrix} \tag{9.9}$$

Similar relationships can also be obtained between phase currents and their sequence components as below:

$$\begin{bmatrix} \boldsymbol{I}_a \\ \boldsymbol{I}_b \\ \boldsymbol{I}_c \end{bmatrix} = \begin{bmatrix} 1 & 1 & 1 \\ 1 & \boldsymbol{a}^2 & \boldsymbol{a} \\ 1 & \boldsymbol{a} & \boldsymbol{a}^2 \end{bmatrix} \times \begin{bmatrix} \boldsymbol{I}_{a0} \\ \boldsymbol{I}_{a1} \\ \boldsymbol{I}_{a2} \end{bmatrix} \tag{9.10}$$

It is possible to extract the sequence components from the above phase values as shown below:

$$\begin{bmatrix} V_{a0} \\ V_{a1} \\ V_{a2} \end{bmatrix} = \left(\frac{1}{3}\right)\begin{bmatrix} 1 & 1 & 1 \\ 1 & a & a^2 \\ 1 & a^2 & a \end{bmatrix} \times \begin{bmatrix} V_a \\ V_b \\ V_c \end{bmatrix} \tag{9.11}$$

$$\begin{bmatrix} I_{a0} \\ I_{a1} \\ I_{a2} \end{bmatrix} = \left(\frac{1}{3}\right)\begin{bmatrix} 1 & 1 & 1 \\ 1 & a & a^2 \\ 1 & a^2 & a \end{bmatrix} \times \begin{bmatrix} I_a \\ I_b \\ I_c \end{bmatrix} \tag{9.12}$$

Example 9.2

Obtain the symmetrical components for the set of unbalanced voltages $V_a = 120\angle -100°\ [\text{kV}]$, $V_b = 150\angle 120°\ [\text{kV}]$ and $V_c = 200\angle -180°\ [\text{kV}]$.

Solution

We utilize equation (9.11) described earlier to obtain symmetrical components of voltages from the set of unbalanced phase voltages.

$$\begin{bmatrix} V_{a0} \\ V_{a1} \\ V_{a2} \end{bmatrix} = \left(\frac{1}{3}\right)\begin{bmatrix} 1 & 1 & 1 \\ 1 & a & a^2 \\ 1 & a^2 & a \end{bmatrix} \times \begin{bmatrix} V_a \\ V_b \\ V_c \end{bmatrix}$$

We therefore have

$$\begin{bmatrix} V_{a0} \\ V_{a1} \\ V_{a2} \end{bmatrix} = \left(\frac{1}{3}\right)\begin{bmatrix} 1 & 1 & 1 \\ 1 & a & a^2 \\ 1 & a^2 & a \end{bmatrix} \times \begin{bmatrix} 120\angle -100° \\ 150\angle 120° \\ 200\angle -180° \end{bmatrix} [\text{kV}]$$

$$= \begin{bmatrix} 98.690\angle 177.730° \\ 24.997\angle -86.818° \\ 123.567\angle -51.816° \end{bmatrix} [\text{kV}]$$

Example 9.3

The symmetrical components of a set of unbalanced three-phase currents are $I_a^0 = 10\angle -120°\ [\text{A}]$, $I_a^1 = 15\angle 100°\ [\text{A}]$ and $I_a^2 = 12\angle -200°\ [\text{A}]$. Obtain the original unbalanced phasors.

Solution

We utilize equation (9.10) described earlier to obtain phase currents from symmetrical components of currents.

$$\begin{bmatrix} \boldsymbol{I}_a \\ \boldsymbol{I}_b \\ \boldsymbol{I}_c \end{bmatrix} = \begin{bmatrix} 1 & 1 & 1 \\ 1 & \boldsymbol{a}^2 & \boldsymbol{a} \\ 1 & \boldsymbol{a} & \boldsymbol{a}^2 \end{bmatrix} \times \begin{bmatrix} \boldsymbol{I}_{a0} \\ \boldsymbol{I}_{a1} \\ \boldsymbol{I}_{a2} \end{bmatrix}$$

We therefore have

$$\begin{bmatrix} \boldsymbol{I}_a \\ \boldsymbol{I}_b \\ \boldsymbol{I}_c \end{bmatrix} = \begin{bmatrix} 1 & 1 & 1 \\ 1 & \boldsymbol{a}^2 & \boldsymbol{a} \\ 1 & \boldsymbol{a} & \boldsymbol{a}^2 \end{bmatrix} \times \begin{bmatrix} 10\angle -120^\circ \\ 15\angle 100^\circ \\ 12\angle -200^\circ \end{bmatrix} [\mathrm{A}]$$

$$= \begin{bmatrix} 21.468\angle 151.583^\circ \\ 27.942\angle -66.417^\circ \\ 12.860\angle -124.576^\circ \end{bmatrix} [\mathrm{A}]$$

9.4.2 Sequence Impedances

In any part of a circuit, the voltage drop occurring due to the flow of current of a certain sequence is dependent upon the impedance of that part of the circuit to that current sequence. The impedance of a circuit can be different for different sequence components of current. The impedance a circuit has when only positive-sequence currents are flowing is called positive-sequence impedance. Similarly, impedance of a circuit when only negative-sequence currents are flowing is called negative-sequence impedance, and impedance offered by the circuit to zero-sequence currents is called zero-sequence impedance [3,38,46].

We consider a general three-phase system shown in Figure 9.9 where $\boldsymbol{Z}_a$, impedance offered by the load to phase a current, has three components $\boldsymbol{Z}_{a0}$, $\boldsymbol{Z}_{a1}$, and $\boldsymbol{Z}_{a2}$, where

$$\boldsymbol{Z}_{a0} = \text{impedance to zero} - \text{sequence current } \boldsymbol{I}_{a0}$$

$$\boldsymbol{Z}_{a1} = \text{impedance to positive} - \text{sequence current } \boldsymbol{I}_{a1}$$

$$\boldsymbol{Z}_{a2} = \text{impedance to negative} - \text{sequence current } \boldsymbol{I}_{a2}$$

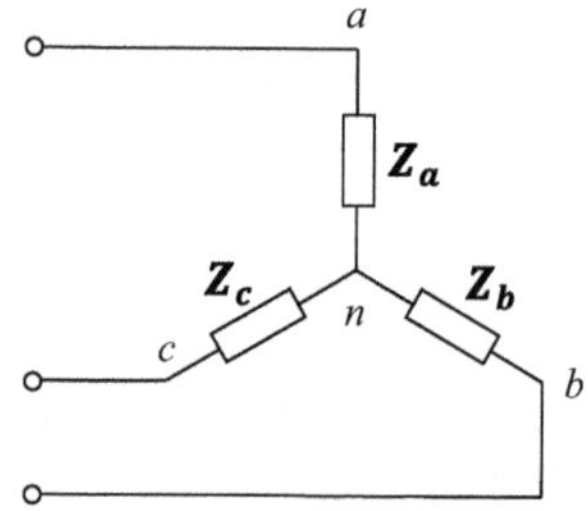

FIGURE 9.9 General three-phase circuit impedance.

Similar quantities can be obtained for impedances of phase b and phase c. We have the following relationships from above:

$$\begin{aligned} V_a &= Z_{a0} \times I_{a0} + Z_{a1} \times I_{a1} + Z_{a2} \times I_{a2} \\ V_b &= Z_{b0} \times I_{b0} + Z_{b1} \times I_{b1} + Z_{b2} \times I_{b2} \\ &= Z_{b0} \times I_{a0} + a^2 Z_{b1} \times I_{a1} + aZ_{b2} \times I_{a2} \\ V_c &= Z_{c0} \times I_{c0} + Z_{c1} \times I_{c1} + Z_{c2} \times I_{c2} \\ &= Z_{c0} \times I_{a0} + aZ_{c1} \times I_{a1} + a^2 Z_{c2} \times I_{a2} \end{aligned} \tag{9.13}$$

We also have

$$\begin{aligned} V_a &= V_{a0} + V_{a1} + V_{a2} \\ V_b &= V_{b0} + V_{b1} + V_{b2} \\ V_c &= V_{c0} + V_{c1} + V_{c2} \end{aligned} \tag{9.14}$$

After equating (9.13) and (9.14), we get

$$\begin{aligned} V_{a0} + V_{a1} + V_{a2} &= Z_{a0} \times I_{a0} + Z_{a1} \times I_{a1} + Z_{a2} \times I_{a2} \\ V_{b0} + V_{b1} + V_{b2} &= Z_{b0} \times I_{a0} + a^2 Z_{b1} \times I_{a1} + aZ_{b2} \times I_{a2} \\ V_{c0} + V_{c1} + V_{c2} &= Z_{c0} \times I_{a0} + aZ_{c1} \times I_{a1} + a^2 Z_{c2} \times I_{a2} \end{aligned} \tag{9.15}$$

$$\begin{aligned} V_{a0} + V_{a1} + V_{a2} &= Z_{a0} \times I_{a0} + Z_{a1} \times I_{a1} + Z_{b2} \times I_{a2} \\ V_{a0} + a^2 V_{a1} + aV_{a2} &= Z_{b0} \times I_{a0} + a^2 Z_{b1} \times I_{a1} + aZ_{b2} \times I_{a2} \\ V_{a0} + aV_{a1} + a^2 V_{a2} &= Z_{c0} \times I_{a0} + aZ_{c1} \times I_{a1} + a^2 Z_{c2} \times I_{a2} \end{aligned} \tag{9.16}$$

In matrix form,

$$\begin{bmatrix} 1 & 1 & 1 \\ 1 & a^2 & a \\ 1 & a & a^2 \end{bmatrix} \times \begin{bmatrix} V_{a0} \\ V_{a1} \\ V_{a2} \end{bmatrix} = \begin{bmatrix} Z_{a0} & Z_{a1} & Z_{a2} \\ Z_{b0} & a^2 Z_{b1} & aZ_{b2} \\ Z_{c0} & aZ_{c1} & a^2 Z_{c2} \end{bmatrix} \times \begin{bmatrix} I_{a0} \\ I_{a1} \\ I_{a2} \end{bmatrix} \tag{9.17}$$

On solving for V_{a0}, V_{a1}, and V_{a2}, we get

$$\begin{bmatrix} V_{a0} \\ V_{a1} \\ V_{a2} \end{bmatrix} = \frac{\begin{bmatrix} Z_{a0} & Z_{a1} & Z_{a2} \\ Z_{b0} & a^2 Z_{b1} & aZ_{b2} \\ Z_{c0} & aZ_{c1} & a^2 Z_{c2} \end{bmatrix}}{\begin{bmatrix} 1 & 1 & 1 \\ 1 & a^2 & a \\ 1 & a & a^2 \end{bmatrix}} \times \begin{bmatrix} I_{a0} \\ I_{a1} \\ I_{a2} \end{bmatrix} \tag{9.18}$$

$$\begin{bmatrix} V_{a0} \\ V_{a1} \\ V_{a2} \end{bmatrix} = \begin{bmatrix} \zeta_{00} & \zeta_{21} & \zeta_{12} \\ \zeta_{10} & \zeta_{01} & \zeta_{22} \\ \zeta_{20} & \zeta_{11} & \zeta_{02} \end{bmatrix} \times \begin{bmatrix} I_{a0} \\ I_{a1} \\ I_{a2} \end{bmatrix} \tag{9.19}$$

where,

$$\zeta_{00} = \frac{Z_{a0} + Z_{b0} + Z_{c0}}{3} \quad \zeta_{21} = \frac{Z_{a1} + a^2 Z_{b1} + aZ_{c1}}{3} \quad \zeta_{12} = \frac{Z_{a2} + aZ_{b2} + a^2 Z_{c2}}{3}$$

$$\zeta_{10} = \frac{Z_{a0} + aZ_{b0} + a^2 Z_{c0}}{3} \quad \zeta_{01} = \frac{Z_{a1} + Z_{b1} + Z_{c1}}{3} \quad \zeta_{22} = \frac{Z_{a2} + a^2 Z_{b2} + aZ_{c2}}{3}$$

$$\zeta_{20} = \frac{Z_{a0} + a^2 Z_{b0} + aZ_{c0}}{3} \quad \zeta_{11} = \frac{Z_{a1} + aZ_{b1} + a^2 Z_{c1}}{3} \quad \zeta_{02} = \frac{Z_{a2} + Z_{b2} + Z_{c2}}{3} \tag{9.20}$$

In equation form, we can write

$$\begin{aligned} V_{a0} &= \zeta_{00} \times I_{a0} + \zeta_{21} \times I_{a1} + \zeta_{12} \times I_{a2} \\ V_{a1} &= \zeta_{10} \times I_{a0} + \zeta_{01} \times I_{a1} + \zeta_{22} \times I_{a2} \\ V_{a2} &= \zeta_{20} \times I_{a0} + \zeta_{11} \times I_{a1} + \zeta_{02} \times I_{a2} \end{aligned} \tag{9.21}$$

We consider, for example, the case of a three-phase rotating machine as a special case characterized by balanced impedances for a particular sequence, and different impedances for different sequence components of current, which means:

$$\begin{aligned} Z_{a0} &= Z_{b0} = Z_{c0} = Z_0 \\ Z_{a1} &= Z_{b1} = Z_{c1} = Z_1 \\ Z_{a2} &= Z_{b2} = Z_{c2} = Z_2 \end{aligned} \tag{9.22}$$

We thus have from equation (9.19) above:

$$\begin{bmatrix} V_{a0} \\ V_{a1} \\ V_{a2} \end{bmatrix} = \begin{bmatrix} Z_0 & 0 & 0 \\ 0 & Z_1 & 0 \\ 0 & 0 & Z_2 \end{bmatrix} \times \begin{bmatrix} I_{a0} \\ I_{a1} \\ I_{a2} \end{bmatrix} \tag{9.23}$$

We can note that the sequence components are independent of one another. Similarly, we can also find that the sequence components of other power system components such as transformers and transmission lines are also independent of each other.

9.5 UNSYMMETRICAL FAULTS ON UNLOADED SYNCHRONOUS GENERATOR

9.5.1 Sequence Networks of an Unloaded Generator

Figure 9.10 shows the circuit diagram of an unloaded generator grounded through a reactance, Z_n. Since the generator circuit supplies balanced three-phase voltages, the generated voltage is of positive sequence only [VII,6,38,46]. As a result, the positive-sequence network consists of the positive-sequence voltage source in series with the positive-sequence impedance, whereas the negative- and zero-sequence networks contain no voltage sources; however, they contain only negative- and zero-sequence impedances, respectively.

Figure 9.11 shows the paths for sequence components of currents that flow through the impedances of their own sequence only, and Figure 9.12 shows the corresponding sequence networks. The sequence networks are equivalent single-phase circuits of the original three-phase circuit through which, during a fault, the sequence components of currents flow. The positive-, negative-, and zero-sequence impedances of the synchronous generator are Z_1, Z_2 and Z_{g0}, respectively.

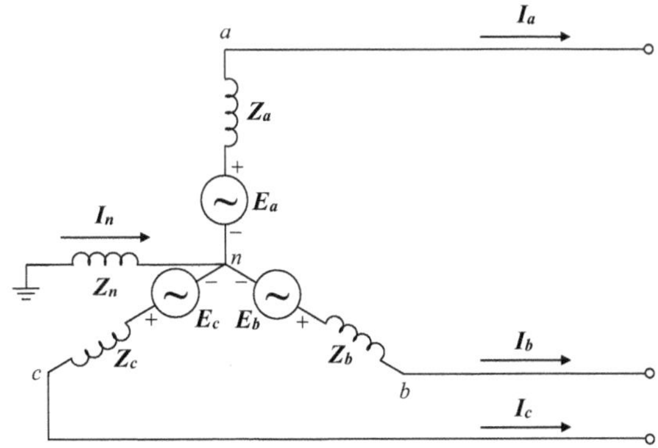

FIGURE 9.10 Unloaded Y-connected generator grounded through an inductive reactance [38].

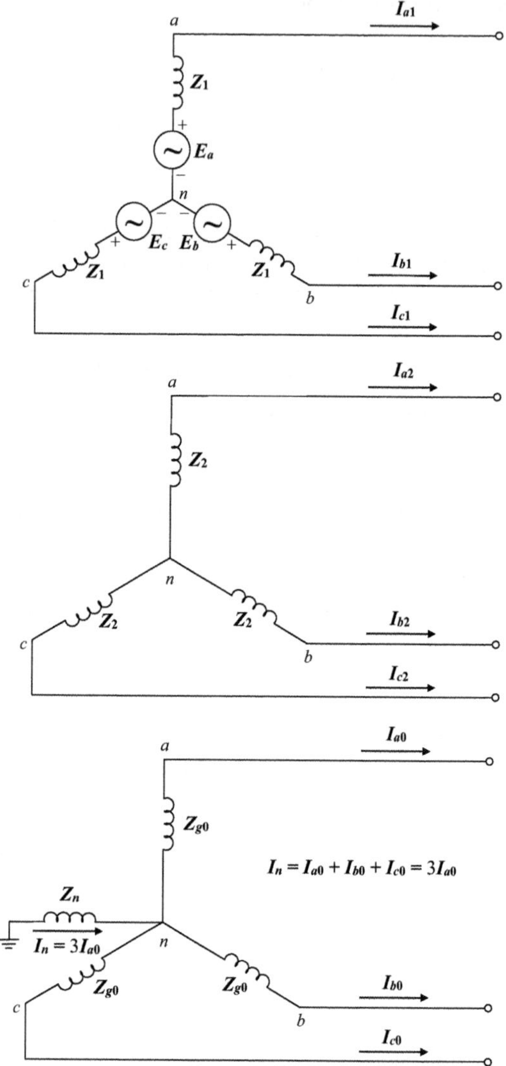

FIGURE 9.11 Paths for the sequence components of currents in a generator [VII].

We have $\mathbf{Z}_0 = \mathbf{Z}_{g0} + 3\mathbf{Z}_n$, since the zero-sequence current through $\mathbf{Z}_n$ is $\mathbf{I}_{a0} + \mathbf{I}_{b0} + \mathbf{I}_{c0} = 3\,\mathbf{I}_{a0}$; however, the current through $\mathbf{Z}_{g0}$ is $\mathbf{I}_{a0}$. In other words, we have

$$\mathbf{V}_{a0} = -\mathbf{I}_{a0} \times \mathbf{Z}_{g0} - 3 \times \mathbf{I}_{a0} \times \mathbf{Z}_n = -\mathbf{I}_{a0} \times \left(\mathbf{Z}_{g0} + 3\mathbf{Z}_n\right) = -\mathbf{I}_{a0} \times \mathbf{Z}_0 \qquad (9.24)$$

$$\mathbf{Z}_0 = \mathbf{Z}_{g0} + 3\mathbf{Z}_n \qquad (9.25)$$

All sequence voltages in sequence network diagrams are measured with respect to the potential reference. The generator neutral is the reference bus for positive- and

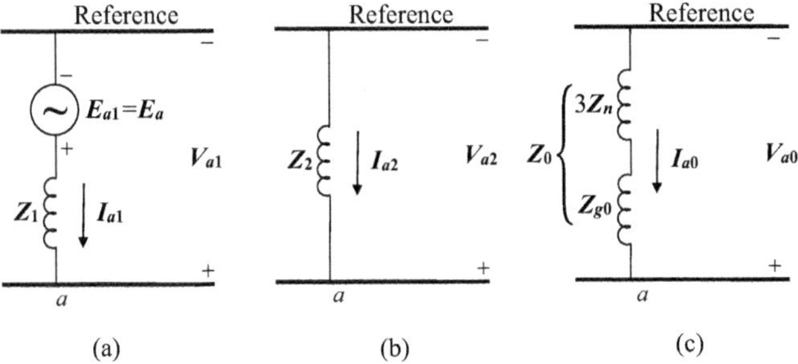

FIGURE 9.12 Positive-, negative-, and zero-sequence networks for an unloaded generator grounded through a reactance [38].

negative-sequence networks since positive- and negative-sequence components represent balanced set, whereas for zero-sequence network, ground at the generator is the reference bus.

From Figure 9.12, we have the following relationships for the voltage and current components of phase a:

$$\begin{aligned} V_{a1} &= E_a - I_{a1} \times Z_1 \\ V_{a2} &= -I_{a2} \times Z_2 \\ V_{a0} &= -I_{a0} \times Z_0 \end{aligned} \tag{9.26}$$

where

E_a = positive-sequence no-load voltage to neutral of the generator
Z_1 = positive-sequence impedance of the generator
Z_2 = negative-sequence impedance of the generator

$$Z_0 = Z_{g0} + 3Z_n$$

Equation (9.26) can also be written in matrix form as below:

$$\begin{bmatrix} V_{a0} \\ V_{a1} \\ V_{a2} \end{bmatrix} = \begin{bmatrix} 0 \\ E_a \\ 0 \end{bmatrix} - \begin{bmatrix} Z_0 & 0 & 0 \\ 0 & Z_1 & 0 \\ 0 & 0 & Z_2 \end{bmatrix} \times \begin{bmatrix} I_{a0} \\ I_{a1} \\ I_{a2} \end{bmatrix} \tag{9.27}$$

There is no mutual coupling between the sequence networks since a particular voltage sequence produces current of that sequence only. We shall next develop the sequence networks for a generator with a fault.

9.5.2 Single Line-to-Ground Fault on an Unloaded Generator

Figure 9.13 shows the circuit for a single line-to-ground fault on an unloaded Y-connected generator which has it neutral grounded through impedance Z_n. The fault occurs on phase a of the generator.

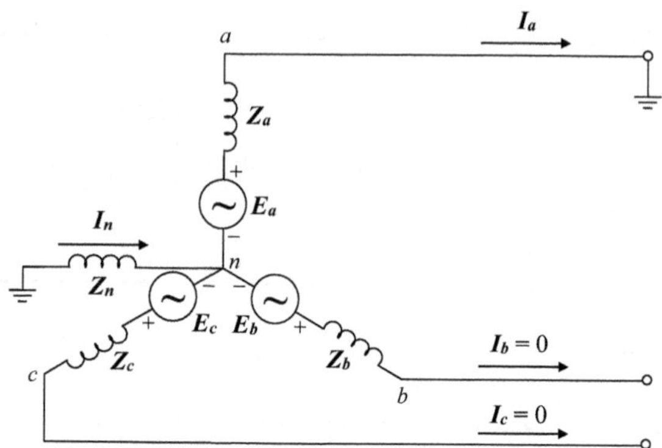

FIGURE 9.13 An unloaded generator with a line-to-ground fault on phase a and neutral grounded through a reactive impedance [VII].

The voltage and current relationships for phase a of this generator are obtained from equations (9.26) described above.

When a single line-to-ground fault occurs on phase a of an unloaded generator, we have the following conditions at the fault:

$$\boldsymbol{I}_b = \boldsymbol{I}_c = 0 \tag{9.28}$$

$$\boldsymbol{V}_a = 0 \tag{9.29}$$

We therefore have

$$\begin{aligned}
\boldsymbol{I}_{a0} &= \frac{1}{3}(\boldsymbol{I}_a + \boldsymbol{I}_b + \boldsymbol{I}_c) = \frac{1}{3}\boldsymbol{I}_a \\
\boldsymbol{I}_{a1} &= \frac{1}{3}\left(\boldsymbol{I}_a + \boldsymbol{a}\boldsymbol{I}_b + \boldsymbol{a}^2\boldsymbol{I}_c\right) = \frac{1}{3}\boldsymbol{I}_a \\
\boldsymbol{I}_{a2} &= \frac{1}{3}\left(\boldsymbol{I}_a + \boldsymbol{a}^2\boldsymbol{I}_b + \boldsymbol{a}\boldsymbol{I}_c\right) = \frac{1}{3}\boldsymbol{I}_a
\end{aligned} \tag{9.30}$$

$$\boldsymbol{I}_{a0} = \boldsymbol{I}_{a1} = \boldsymbol{I}_{a2} = \frac{1}{3}\boldsymbol{I}_a \tag{9.31}$$

Substituting $\boldsymbol{I}_{a1}$ for $\boldsymbol{I}_{a0}$ and $\boldsymbol{I}_{a2}$ in (9.27), we obtain

$$\begin{bmatrix} \boldsymbol{V}_{a0} \\ \boldsymbol{V}_{a1} \\ \boldsymbol{V}_{a2} \end{bmatrix} = \begin{bmatrix} 0 \\ \boldsymbol{E}_a \\ 0 \end{bmatrix} - \begin{bmatrix} \boldsymbol{Z}_0 & 0 & 0 \\ 0 & \boldsymbol{Z}_1 & 0 \\ 0 & 0 & \boldsymbol{Z}_2 \end{bmatrix} \times \begin{bmatrix} \boldsymbol{I}_{a1} \\ \boldsymbol{I}_{a1} \\ \boldsymbol{I}_{a1} \end{bmatrix} \tag{9.32}$$

$$V_{a0} + V_{a1} + V_{a2} = (E_a - I_{a1} \times Z_1 - I_{a1} \times Z_2 - I_{a1} \times Z_0) \tag{9.33}$$

Since $V_a = V_{a0} + V_{a1} + V_{a2} = 0$, we have

$$I_{a1} = \frac{E_a}{Z_1 + Z_2 + Z_0} \tag{9.34}$$

$$I_a = \frac{3E_a}{Z_1 + Z_2 + Z_0} \tag{9.35}$$

where

$$Z_0 = Z_{g0} + 3Z_n \tag{9.36}$$

The above equations describe a series connection of the three-sequence networks as shown in Figure 9.14 [VII,6,38,46]. The equations obtained above for single line-to-ground fault are used with equation set (9.26) and the symmetrical component relations described earlier to determine all the voltages and currents at the fault.

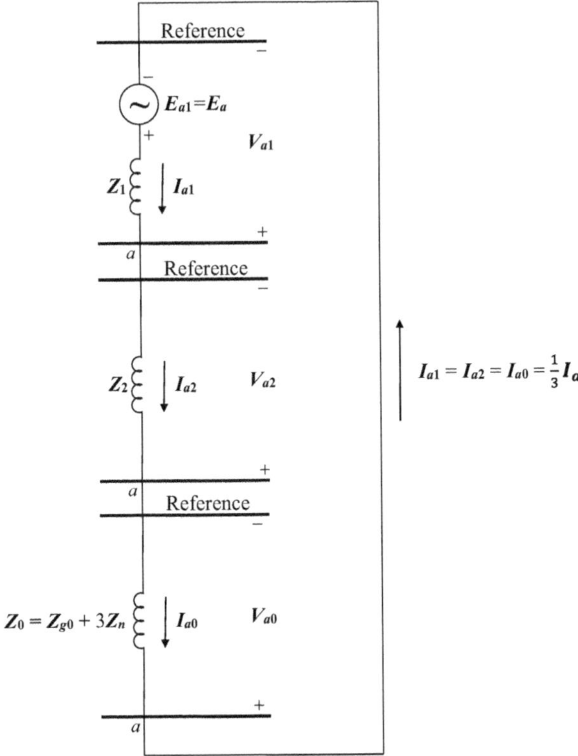

FIGURE 9.14 Sequence network connections for an unloaded generator with a single line-to-ground fault on phase *a* [38].

9.5.3 Line-to-Line Fault on an Unloaded Generator

Figure 9.15 shows the circuit for line-to-line fault on an unloaded Y-connected generator which has it neutral grounded through impedance $\boldsymbol{Z_n}$. The fault occurs between phases b and c of the generator.

When a line-to-line fault occurs between phases b and c of an unloaded generator, we have the following conditions at the fault:

$$\boldsymbol{I_a} = 0 \tag{9.37}$$

$$\boldsymbol{I_b} = -\boldsymbol{I_c} \tag{9.38}$$

$$\boldsymbol{V_b} = \boldsymbol{V_c} \tag{9.39}$$

Since $\boldsymbol{V_b} = \boldsymbol{V_c}$, the symmetrical components of voltage from equation (9.11) are

$$\begin{bmatrix} \boldsymbol{V_{a0}} \\ \boldsymbol{V_{a1}} \\ \boldsymbol{V_{a2}} \end{bmatrix} = \left(\frac{1}{3}\right) \begin{bmatrix} 1 & 1 & 1 \\ 1 & \boldsymbol{a} & \boldsymbol{a}^2 \\ 1 & \boldsymbol{a}^2 & \boldsymbol{a} \end{bmatrix} \times \begin{bmatrix} \boldsymbol{V_a} \\ \boldsymbol{V_b} \\ \boldsymbol{V_b} \end{bmatrix} \tag{9.40}$$

From the above set of equations, we obtain

$$\boldsymbol{V_{a1}} = \boldsymbol{V_{a2}} \tag{9.41}$$

We also have $\boldsymbol{I_a} = 0$ and $\boldsymbol{I_b} = -\boldsymbol{I_c}$. The symmetrical components of current from equation (9.12) therefore are

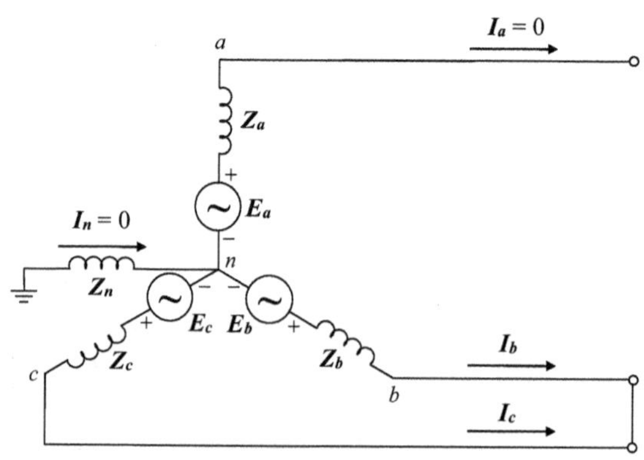

FIGURE 9.15 An unloaded generator with a line-to-line fault between phases b and c, and neutral grounded through a reactive impedance [VII].

$$\begin{bmatrix} I_{a0} \\ I_{a1} \\ I_{a2} \end{bmatrix} = \left(\frac{1}{3}\right) \begin{bmatrix} 1 & 1 & 1 \\ 1 & a & a^2 \\ 1 & a^2 & a \end{bmatrix} \times \begin{bmatrix} I_a \\ -I_c \\ I_c \end{bmatrix} \tag{9.42}$$

We thus have

$$I_{a0} = 0 \tag{9.43}$$

$$I_{a2} = -I_{a1} \tag{9.44}$$

Since $I_{a0} = 0$, we obtain

$$V_{a0} = -I_{a0} \times Z_0 = -0 \times Z_0 = 0$$

$$V_{a0} = 0 \tag{9.45}$$

Substituting relationships obtained above for V_{a0}, V_{a1}, V_{a2}, I_{a0}, I_{a1} and I_{a2} in equation (9.27), we obtain

$$\begin{bmatrix} 0 \\ V_{a1} \\ V_{a1} \end{bmatrix} = \begin{bmatrix} 0 \\ E_a \\ 0 \end{bmatrix} - \begin{bmatrix} Z_0 & 0 & 0 \\ 0 & Z_1 & 0 \\ 0 & 0 & Z_2 \end{bmatrix} \times \begin{bmatrix} 0 \\ I_{a1} \\ -I_{a1} \end{bmatrix} \tag{9.46}$$

We therefore have

$$\begin{aligned} V_{a1} &= E_a - I_{a1} \times Z_1 \\ V_{a1} &= I_{a1} \times Z_2 \\ 0 &= -\ 0 \times Z_0 \end{aligned} \tag{9.26}$$

From the above equations, we obtain

$$\begin{aligned} I_{a1} \times Z_2 &= E_a - I_{a1} \times Z_1 \\ I_{a1} &= \frac{E_a}{Z_1 + Z_2} \end{aligned} \tag{9.47}$$

The results obtained above for line-to-line fault require the connection of positive- and negative-sequence networks in parallel [VII,6,38,46]. There is no zero-sequence network since Z_0 is absent in the above equations. The connection of the sequence networks for a line-to-line fault is shown in Figure 9.16. The equations obtained above for line-to-line fault are used with equation set (9.26) and the symmetrical component relations described earlier to determine all the voltages and currents at the fault.

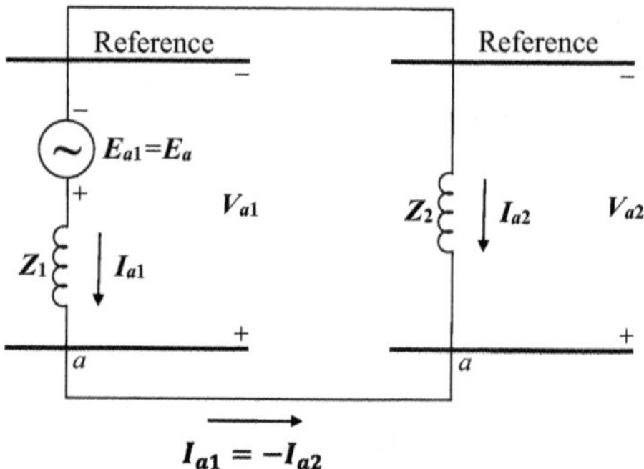

FIGURE 9.16 Connection of the positive- and negative-sequence networks for a line-to-line fault between phases b and c of the unloaded generator [VII].

9.5.4 Double Line-to-Ground Fault on an Unloaded Generator

Figure 9.17 shows the circuit for double line-to-ground fault on an unloaded Y-connected generator which has it neutral grounded through impedance $\boldsymbol{Z_n}$. The phases of the generator faulted to ground are b and c.

When a double line-to-ground fault occurs between phases b and c, and ground of an unloaded generator, we have the following conditions at the fault:

$$V_b = 0 \tag{9.48}$$

$$V_c = 0 \tag{9.49}$$

$$I_a = 0 \tag{9.50}$$

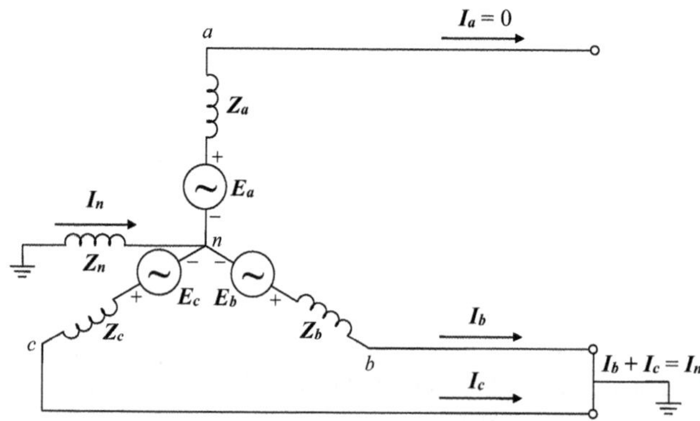

FIGURE 9.17 An unloaded generator with a double line-to-ground fault between phases b and c, and ground, and with the neutral grounded through a reactive impedance [38].

Since $V_b = V_c = 0$, the symmetrical components of voltage from equation (9.11) are

$$\begin{bmatrix} V_{a0} \\ V_{a1} \\ V_{a2} \end{bmatrix} = \left(\frac{1}{3}\right)\begin{bmatrix} 1 & 1 & 1 \\ 1 & a & a^2 \\ 1 & a^2 & a \end{bmatrix} \times \begin{bmatrix} V_a \\ 0 \\ 0 \end{bmatrix} \tag{9.51}$$

We therefore have

$$V_{a1} = V_{a2} = V_{a0} = V_a / 3 \tag{9.52}$$

We substitute $E_a - I_{a1} \times Z_1$ for V_{a0}, V_{a1} and V_{a2} in equation (9.27) and obtain

$$\begin{bmatrix} E_a - I_{a1} \times Z_1 \\ E_a - I_{a1} \times Z_1 \\ E_a - I_{a1} \times Z_1 \end{bmatrix} = \begin{bmatrix} 0 \\ E_a \\ 0 \end{bmatrix} - \begin{bmatrix} Z_0 & 0 & 0 \\ 0 & Z_1 & 0 \\ 0 & 0 & Z_2 \end{bmatrix} \times \begin{bmatrix} I_{a0} \\ I_{a1} \\ I_{a2} \end{bmatrix} \tag{9.53}$$

We next pre-multiply both sides of the above matrix equation with

$$\begin{bmatrix} Z_0 & 0 & 0 \\ 0 & Z_1 & 0 \\ 0 & 0 & Z_2 \end{bmatrix}^{-1} = \begin{bmatrix} \frac{1}{Z_0} & 0 & 0 \\ 0 & \frac{1}{Z_1} & 0 \\ 0 & 0 & \frac{1}{Z_2} \end{bmatrix}$$

We then obtain

$$\begin{bmatrix} \frac{1}{Z_0} & 0 & 0 \\ 0 & \frac{1}{Z_1} & 0 \\ 0 & 0 & \frac{1}{Z_2} \end{bmatrix} \begin{bmatrix} E_a - I_{a1} \times Z_1 \\ E_a - I_{a1} \times Z_1 \\ E_a - I_{a1} \times Z_1 \end{bmatrix} = \begin{bmatrix} \frac{1}{Z_0} & 0 & 0 \\ 0 & \frac{1}{Z_1} & 0 \\ 0 & 0 & \frac{1}{Z_2} \end{bmatrix} \begin{bmatrix} 0 \\ E_a \\ 0 \end{bmatrix} - \begin{bmatrix} I_{a0} \\ I_{a1} \\ I_{a2} \end{bmatrix}$$

$$\begin{bmatrix} \frac{1}{Z_0}(E_a - I_{a1} \times Z_1) \\ \frac{1}{Z_1}(E_a - I_{a1} \times Z_1) \\ \frac{1}{Z_2}(E_a - I_{a1} \times Z_1) \end{bmatrix} = \begin{bmatrix} 0 \\ \frac{E_a}{Z_1} \\ 0 \end{bmatrix} - \begin{bmatrix} I_{a0} \\ I_{a1} \\ I_{a2} \end{bmatrix} \tag{9.54}$$

We thus have from above

$$\frac{1}{Z_0}(E_a - I_{a1} \times Z_1) = -I_{a0}$$

$$\frac{1}{Z_1}(E_a - I_{a1} \times Z_1) = \frac{E_a}{Z_1} - I_{a1} \tag{9.55}$$

$$\frac{1}{Z_2}(E_a - I_{a1} \times Z_1) = -I_{a2}$$

After summing the above equations, we obtain

$$\frac{1}{Z_0}(E_a - I_{a1} \times Z_1) + \frac{1}{Z_1}(E_a - I_{a1} \times Z_1) + \frac{1}{Z_2}(E_a - I_{a1} \times Z_1) = \frac{E_a}{Z_1}$$

$$\frac{E_a}{Z_0} - I_{a1}\frac{Z_1}{Z_0} + \frac{E_a}{Z_1} - I_{a1} + \frac{E_a}{Z_2} - I_{a1}\frac{Z_1}{Z_2} = \frac{E_a}{Z_1}$$

$$I_{a1}\left(1 + \frac{Z_1}{Z_0} + \frac{Z_1}{Z_2}\right) = \frac{E_a(Z_2 + Z_0)}{Z_2 Z_0}$$

$$I_{a1} = \frac{E_a(Z_2 + Z_0)}{Z_1 Z_2 + Z_1 Z_0 + Z_2 Z_0} = \frac{E_a}{Z_1 + Z_2 Z_0/(Z_2 + Z_0)} \tag{9.56}$$

The results obtained above for double line-to-ground require connection of positive-, negative-, and zero-sequence networks in parallel [VII,6,38,46]. The connection of the sequence networks for a line-to-line fault is shown in Figure 9.18. The equations obtained above for double line-to-ground fault are used with equation set (9.26) and the symmetrical component relations described earlier to determine all the voltages and currents at the fault.

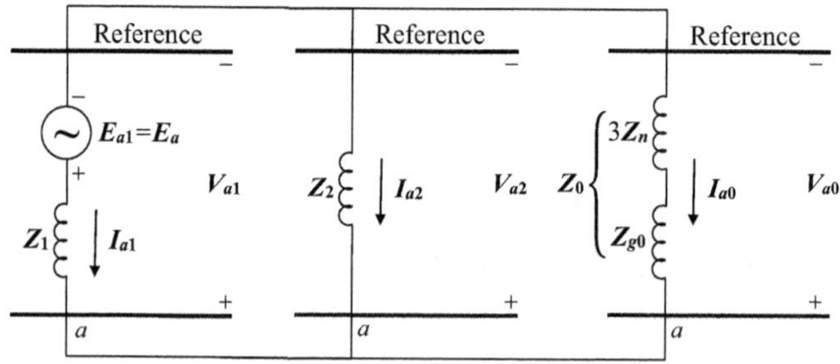

FIGURE 9.18 Connection of positive-, negative-, and zero-sequence networks for a double line-to-ground fault between phases *b* and *c*, and ground of the unloaded generator [38].

Example 9.4

A Y-connected synchronous generator has a positive-sequence reactance of 0.25 [pu], a negative-sequence reactance of 0.35 [pu], and a zero-sequence reactance of 0.15 [pu]. The generator neutral is solidly grounded. The generator is initially unloaded and operated at rated voltage of 1 [pu], and a single line-to-ground fault occurs on phase *a* at the terminals of the generator with fault location labeled F. Find the currents and voltages in each phase of the generator after the fault occurs.

Solution

The machine is initially operating at no load and rated voltage. We therefore have $\boldsymbol{E}_a = 1.0\angle 0°[\text{pu}]$. The sequence networks for this generator are shown in Figure 9.19, with $\mathbf{Z}_1 = j0.25\,[\text{pu}]$, $\mathbf{Z}_2 = j0.35\,[\text{pu}]$, and $\mathbf{Z}_0 = j0.15\,[\text{pu}]$. The resulting sequence currents in phase *a* are:

$$\boldsymbol{I}_{a0} = \boldsymbol{I}_{a1} = \boldsymbol{I}_{a2} = \frac{\boldsymbol{E}_a}{\mathbf{Z}_0 + \mathbf{Z}_1 + \mathbf{Z}_2}$$

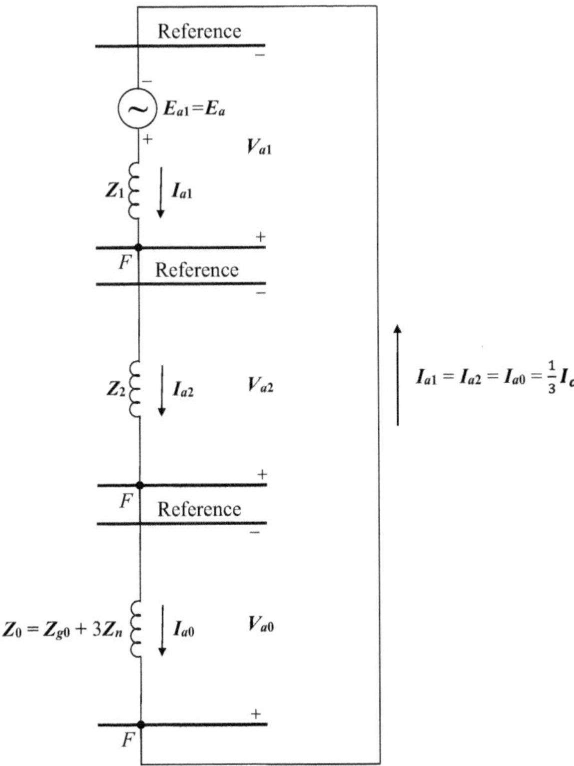

FIGURE 9.19 Connection of the positive-, negative-, and zero-sequence networks for a single line-to-ground fault at the terminals of synchronous generator (fault location represented by letter F) to determine the sequence currents in the fault.

$$\boldsymbol{I}_{a0} = \boldsymbol{I}_{a1} = \boldsymbol{I}_{a2} = \frac{1\angle 0^\circ}{j0.25 + j0.35 + j0.15} = 1.333\angle -90^\circ \text{ [pu]}$$

We thus have:

$$\begin{aligned} \boldsymbol{I}_a &= \boldsymbol{I}_{a0} + \boldsymbol{I}_{a1} + \boldsymbol{I}_{a2} \\ &= 1.333\angle -90^\circ + 1.333\angle -90^\circ + 1.333\angle -90^\circ \\ &= 4\angle -90^\circ \text{ [pu]} \end{aligned}$$

$$\begin{aligned} \boldsymbol{I}_b &= \boldsymbol{I}_{a0} + \boldsymbol{a}^2 \boldsymbol{I}_{a1} + \boldsymbol{a}\boldsymbol{I}_{a2} \\ &= (1.333\angle -90^\circ) + \boldsymbol{a}^2(1.333\angle -90^\circ) + \boldsymbol{a}(1.333\angle -90^\circ) \\ &= 0.0 \text{ [pu]} \end{aligned}$$

$$\begin{aligned} \boldsymbol{I}_c &= \boldsymbol{I}_{a0} + \boldsymbol{a}\boldsymbol{I}_{a1} + \boldsymbol{a}^2 \boldsymbol{I}_{a2} \\ &= (1.333\angle -90^\circ) + \boldsymbol{a}(1.333\angle -90^\circ) + \boldsymbol{a}^2(1.333\angle -90^\circ) \\ &= 0.0 \text{ [pu]} \end{aligned}$$

The phase currents in phases *b* and *c* are zero, since they are open-circuited.

The sequence voltages in phase *a* are:

$$\begin{aligned} \boldsymbol{V}_{a1} &= \boldsymbol{E}_{a1} - \boldsymbol{I}_{a1} \times \boldsymbol{Z}_1 \\ &= 1\angle 0^\circ - (1.333\angle -90^\circ)(j0.25) = 0.667\angle 0^\circ \text{ [pu]} \end{aligned}$$

$$\begin{aligned} \boldsymbol{V}_{a2} &= -\boldsymbol{I}_{a2} \times \boldsymbol{Z}_2 \\ &= -(1.333\angle -90^\circ)(j0.35) = 0.467\angle 180^\circ \text{ [pu]} \end{aligned}$$

$$\begin{aligned} \boldsymbol{V}_{a0} &= -\boldsymbol{I}_{a0} \times \boldsymbol{Z}_0 \\ &= -(1.333\angle -90^\circ)(j0.15) = 0.200\angle 180^\circ \text{ [pu]} \end{aligned}$$

We thus have:

$$\begin{aligned} \boldsymbol{V}_a &= \boldsymbol{V}_{a0} + \boldsymbol{V}_{a1} + \boldsymbol{V}_{a2} \\ &= (0.200\angle 180^\circ) + (0.667\angle 0^\circ) + (0.467\angle 180^\circ) \\ &= 0.0\angle 0^\circ \text{ [pu]} \end{aligned}$$

$$\begin{aligned} V_b &= V_{a0} + a^2 V_{a1} + a V_{a2} \\ &= (0.200\angle 180°) + a^2(0.667\angle 0°) + a(0.467\angle 180°) \\ &= 1.027\angle -106.987° \text{ [pu]} \end{aligned}$$

$$\begin{aligned} V_c &= V_{a0} + a V_{a1} + a^2 V_{a2} \\ &= (0.200\angle 180°) + a(0.667\angle 0°) + a^2(0.467\angle 180°) \\ &= 1.027\angle 106.987° \text{ [pu]} \end{aligned}$$

The generator terminal phase *a* voltage is zero, since it is short circuited.

9.6 UNSYMMETRICAL FAULTS ON POWER SYSTEMS

The previous section described calculation of voltages and currents during various unsymmetrical faults on an unloaded synchronous generator. This description is expanded in present section to illustrate the calculation of voltages and currents during various unsymmetrical faults on larger power systems [VII,6,46].

We consider the power system shown in Figure 9.20. It is assumed that a fault occurs at location "×". Figure 9.21 shows the positive-, negative-, and zero-sequence networks for this power system together with their Thevenin equivalent circuits. The location of the fault is indicated on each sequence network. The procedure for drawing the sequence networks for a loaded generator or power system is similar to that for an unloaded generator described earlier. However, when transformers are present, all system impedances are expressed in per unit in order to eliminate the necessity of voltage and current transformations at different voltage levels. While drawing the zero-sequence network, we also keep in mind the following information:

1. An ungrounded wye connection presents infinite impedance to zero-sequence current.
2. If the neutral point of a wye is grounded through an impedance $\mathbf{Z}_n$, $3\mathbf{Z}_n$ is placed between the neutral location and ground.
3. A Δ-connected circuit offers infinite impedance to zero-sequence line currents. The zero-sequence currents can however circulate in the Δ circuit.

The Thevenin equivalent circuits between fault location and reference bus location shown for each sequence network are also drawn in Figure 9.21. We find that the Thevenin equivalent circuits of positive-, negative- and zero-sequence networks

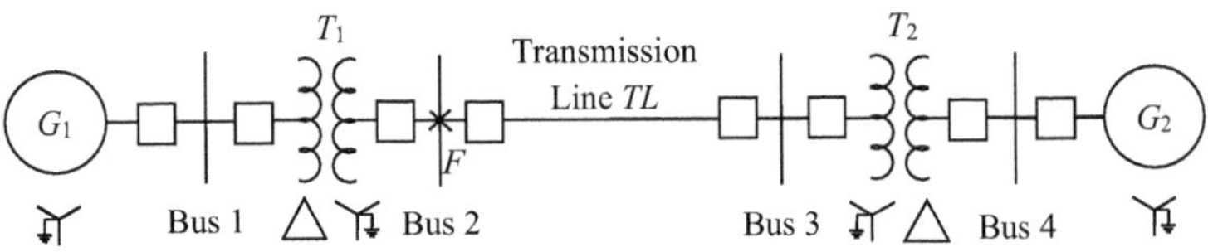

FIGURE 9.20 Simple power system with a fault at the location indicated by ×.

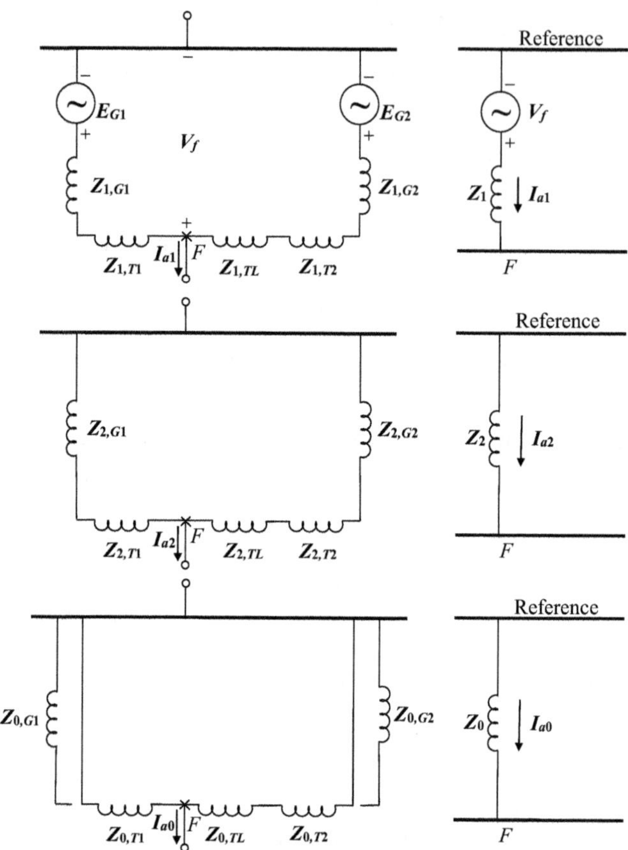

FIGURE 9.21 Positive-, negative-, and zero-sequence networks for the power system together with their Thevenin equivalent circuits in which V_f is the fault location pre-fault voltage.

of the power system are exactly similar to corresponding sequence networks of an unloaded synchronous generator. The description in previous section about unsymmetrical faults on an unloaded synchronous generator will therefore apply to unsymmetrical faults on power systems if the power system sequence networks are represented by their corresponding Thevenin equivalent circuits.

9.6.1 Single Line-to-Ground Fault on a Power System

For a single line-to-ground fault on phase *a* of a power system as shown in Figure 9.22, we have the following conditions at the fault:

$$I_b = I_c = 0 \tag{9.57}$$

$$V_a = 0 \tag{9.58}$$

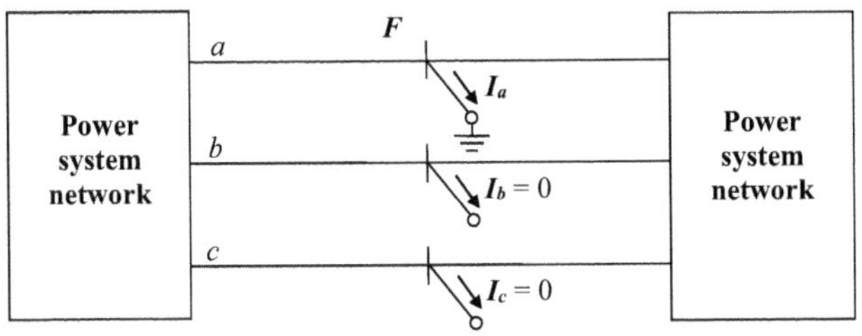

FIGURE 9.22 Single line-to-ground fault on a power system [VII].

The above relationships for single line-to-ground fault on a power system are exactly similar to those for single line-to-ground fault on an unloaded generator. The equations for determining symmetrical components of fault voltage and fault current in a power system are therefore also similar to those obtained earlier for unloaded synchronous generator, with the only difference that we replace $\boldsymbol{E}_a$ with fault location pre-fault voltage $\boldsymbol{V}_f$. We therefore have the following equation for calculating $\boldsymbol{I}_{a1}$:

$$\boldsymbol{I}_{a1} = \frac{\boldsymbol{V}_f}{\boldsymbol{Z}_0 + \boldsymbol{Z}_1 + \boldsymbol{Z}_2} \tag{9.59}$$

In the above equation, $\boldsymbol{V}_f$ is pre-fault voltage at the fault location, and $\boldsymbol{Z}_0$, $\boldsymbol{Z}_1$ and $\boldsymbol{Z}_2$ are Thevenin equivalent circuit impedances of the power system sequence networks between two terminals consisting of the fault location and reference bus. As the above equation suggests, the three-sequence networks should be connected in series at the fault location F to determine fault current and other power system quantities during single line-to-ground fault [VII,6,46]. Figure 9.23 shows the connection of sequence networks for a single line-to-ground fault on a power system.

9.6.2 Line-to-Line Fault on a Power System

For a line-to-line fault between phases b and c of a power system as shown in Figure 9.24, we have the following conditions at the fault:

$$\boldsymbol{I}_a = 0 \tag{9.60}$$

$$\boldsymbol{I}_b = -\boldsymbol{I}_c \tag{9.61}$$

$$\boldsymbol{V}_b = \boldsymbol{V}_c \tag{9.62}$$

The above relationships for a line-to-line fault on a power system are exactly similar to those for a line-to-line fault on an unloaded generator. The equations for determining symmetrical components of fault voltage and fault current in a power system are therefore also similar to those obtained earlier for unloaded synchronous generator, with the only difference that we replace $\boldsymbol{E}_a$ with fault location pre-fault voltage $\boldsymbol{V}_f$. We therefore have the following equation for calculating $\boldsymbol{I}_{a1}$:

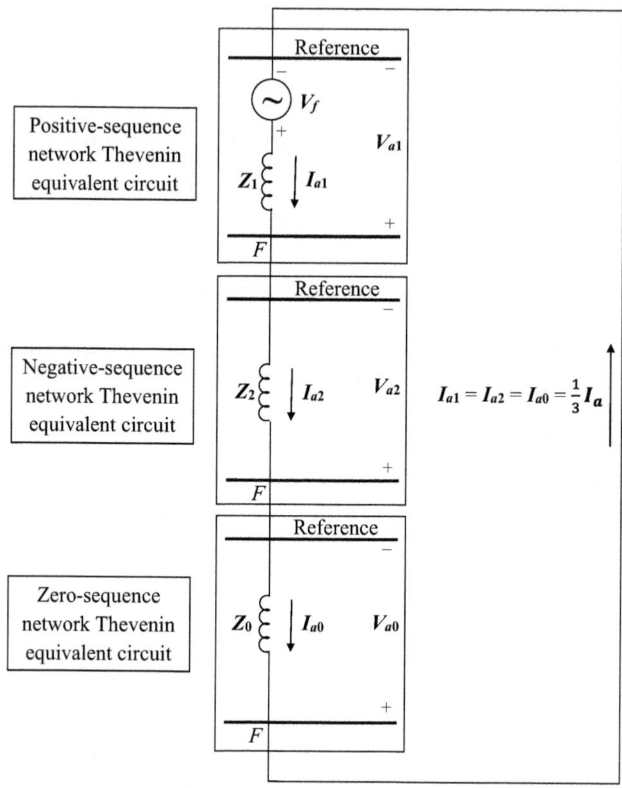

FIGURE 9.23 Connection of the positive-, negative-, and zero-sequence networks for a single line-to-ground fault on a power system at the fault location *F* to determine the sequence voltages and currents [VII].

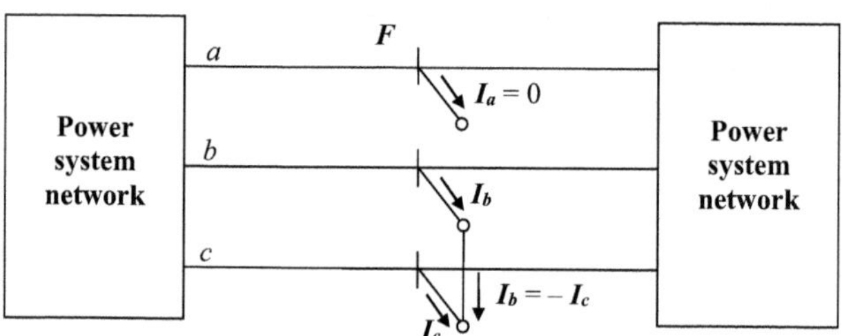

FIGURE 9.24 Line-to-line fault on a power system [VII].

$$I_{a1} = \frac{V_f}{Z_1 + Z_2} \tag{9.63}$$

In the above equation, V_f is pre-fault voltage at the fault location, and Z_1 and Z_2 are Thevenin equivalent circuit impedances of the power system sequence networks between two terminals consisting of the fault location and reference bus. As the above equation suggests, the positive- and negative-sequence networks should be connected in parallel at the fault location F to determine fault current and other power system quantities during a line-to-line fault [VII,6,46]. Figure 9.25 shows the connection of sequence networks for a line-to-line fault on a power system.

9.6.3 Double Line-to-Ground Fault on a Power System

For a double line-to-ground fault on a power system with phases b and c faulted to ground as shown in Figure 9.26, we have the following conditions at the fault:

$$V_b = 0 \tag{9.64}$$

$$V_c = 0 \tag{9.65}$$

$$I_a = 0 \tag{9.66}$$

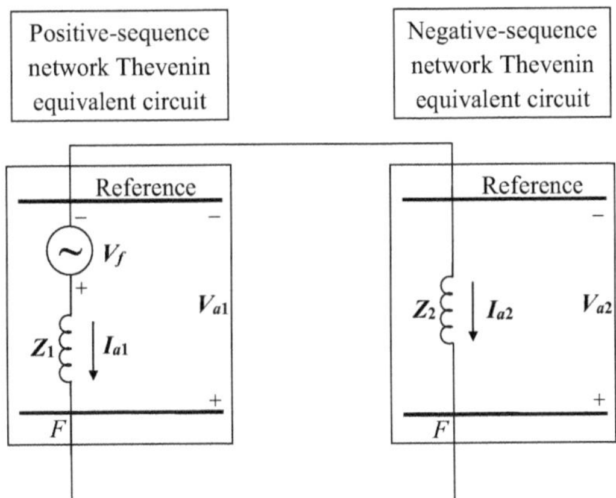

FIGURE 9.25 Connection of positive- and negative-sequence networks for a line-to-line fault on a power system at the fault location F to determine the sequence voltages and currents [VII].

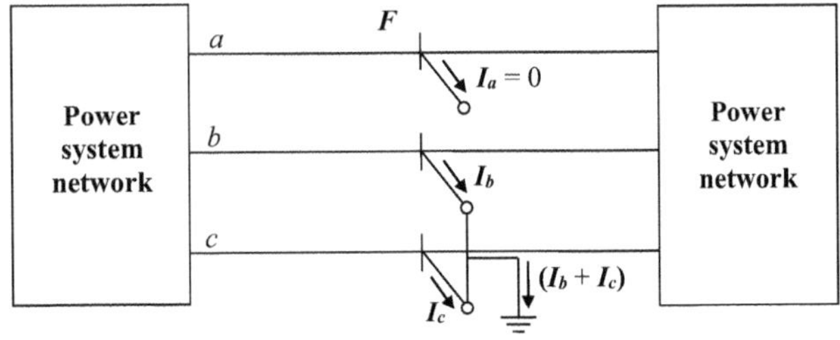

FIGURE 9.26 Double line-to-ground fault on a power system [VII].

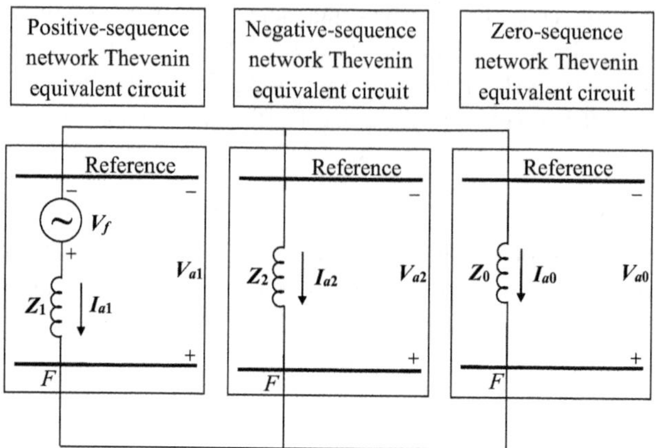

FIGURE 9.27 Connection of the positive-, negative-, and zero-sequence networks for double line-to-ground fault on a power system at the fault location F to determine the sequence voltages and currents [VII].

The above relationships for a double line-to-ground fault on a power system are exactly similar to those for a double line-to-ground fault on an unloaded generator. The equations for determining symmetrical components of fault voltage and fault current in a power system are therefore also similar to those obtained earlier for unloaded synchronous generator, with the only difference that we replace $\boldsymbol{E_a}$ with fault location pre-fault voltage $\boldsymbol{V_f}$. We therefore have the following equation for calculating $\boldsymbol{I_{a1}}$:

$$\boldsymbol{I_{a1}} = \frac{\boldsymbol{V_f}}{\boldsymbol{Z_1} + \dfrac{\boldsymbol{Z_2 Z_0}}{(\boldsymbol{Z_2} + \boldsymbol{Z_0})}} \tag{9.67}$$

In the above equation, $\boldsymbol{V_f}$ is pre-fault voltage at the fault location, and $\boldsymbol{Z_0}$, $\boldsymbol{Z_1}$ and $\boldsymbol{Z_2}$ are Thevenin equivalent circuit impedances of the power system sequence networks between two terminals consisting of the fault location and reference bus. As the above equation suggests, the three-sequence networks should be connected in parallel at the fault location F to determine fault current and other power system quantities during a double line-to-ground fault [VII,6,46]. Figure 9.27 shows the connection of sequence networks for a double line-to-ground fault on a power system.

Example 9.5

The reactance data for the power system shown in Figure 9.28 in per unit on a common base is shown in Table 9.1 below.

The pre-fault voltage $\boldsymbol{V_f}$ at the location of the fault is 1 per unit. Draw the sequence impedance networks for the fault at bus 1 and compute the fault current in per unit for a bolted single line-to-ground fault at bus 1.

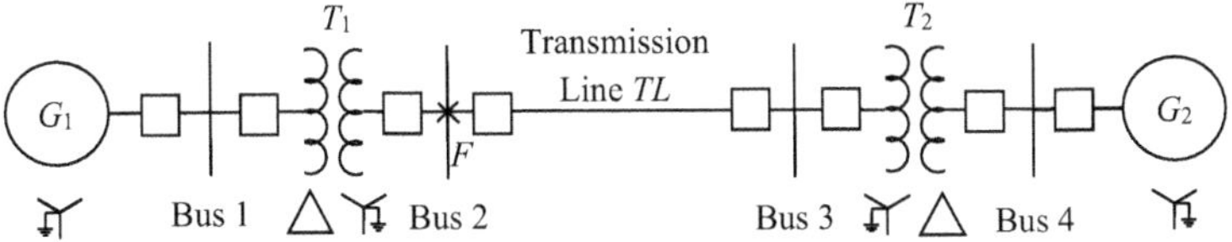

FIGURE 9.28 Example power system.

TABLE 9.1
Reactance Data for the Example Power System

Component	X_1	X_2	X_0
G_1	0.15	0.15	0.10
G_2	0.15	0.15	0.10
T_1	0.35	0.35	0.35
T_2	0.35	0.35	0.35
TL	0.25	0.25	0.55

Solution

The positive-sequence impedance network obtained after ignoring the voltage sources is shown in Figure 9.29, and impedance to the location of fault is:

$$\mathbf{Z}_1 = j\frac{(0.50)(0.75)}{0.50+0.75} = j0.3\ [\text{pu}]$$

The negative-sequence reactances are the same as positive-sequence reactances, and we thus have $X_2 = X_1 = 0.3$ per unit. The zero-sequence impedance network is shown in Figure 9.30, and impedance to the location of fault is:

$$\mathbf{Z}_0 = j\frac{(0.35)(0.90)}{0.35+0.90} = j0.252\ [\text{pu}]$$

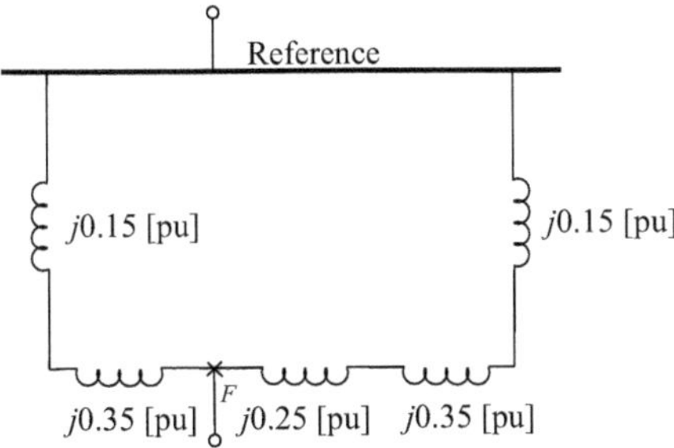

FIGURE 9.29 Positive-sequence impedance network after ignoring the voltage sources for the example power system.

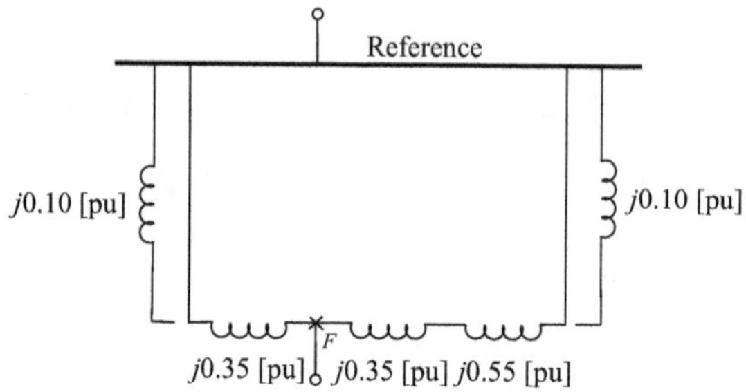

FIGURE 9.30 Zero-sequence impedance network for the example power system.

We have:

$$I_{a0} = \frac{1\angle 0^\circ}{j(0.3+0.3+0.252)} = 1.174\angle -90^\circ \ [\text{pu}]$$

For a bolted single line-to-ground fault at bus 1, the fault current is:

$$I_f = 3I_{a0} = 3.521\angle -90^\circ \ [\text{pu}]$$

The current flowing in any particular power system component such as the transmission line can be obtained by summing positive-sequence, negative-sequence and zero-sequence components of current through it using the original sequence network diagram. The student reader is requested to refer to standard textbooks mentioned in the list of references for more detailed description of fault analysis of Electric Power Systems.

Part III: Power System Protection

9.7 INTRODUCTION TO POWER SYSTEM PROTECTION

Faults occur on even the best designed electrical systems. Power system faults result in abnormally high currents in the power conductors or in the ground conductors, or both. The consequence of not removing a fault is damage to faulted as well as unfaulted parts of the power system. The purpose of a protection system is to sense the fault condition and disconnect the malfunctioning equipment or section as quickly and as selectively as possible, thus preventing or minimizing harm to equipment and personnel.

The purpose of this description is to provide an introduction to power system protection. The main components of the protection system along with relaying schemes available for the protection of generators, transformers and transmission lines are briefly described below. Because of its relevance to power system study, power system stability is also briefly described in the end.

9.8 PROTECTION SYSTEM COMPONENTS

The purpose of the protection system is to detect and remove the faulted equipment or section of the power system as fast and as selectively as possible. A well-designed protection system should be:

- reliable (capable of detecting all types of faults),
- selective (isolating only the faulted part of the system),
- fast (operating rapidly to minimize fault duration and equipment damage),
- sensitive (ability to operate with low value of actuating quantity),
- economical (providing maximum protection at minimum cost), and
- simple (requiring minimum amount of equipment and circuitry).

The main components of a protection systems are:

1. Transducers (instrument transformers): sensor,
2. Relays: decision maker,
3. Circuit breakers: offending circuit de-energizer.

The three components above can be present as separate devices in a protection system, or as different functions within a single device. In this description, transducers, relays and circuit breakers represent the functions of sensing, relaying and circuit de-energizing respectively performed by them irrespective of the form in which they are present in a protection system.

Figure 9.31 is provided to help in the understanding of usefulness of different components mentioned above in power system protection. It shows a simple over-current protection scheme consisting of a current transformer, an over-current relay, and a circuit breaker. Only one phase of three-phase system is shown for simplified representation. After the occurrence of fault on the power transmission line, the current flowing in the line increases considerably. As a result, large amount of current flows through the current transformer and relay coil. The relay then operates by closing its normally open contacts. This in turn results in energizing of trip coil of the

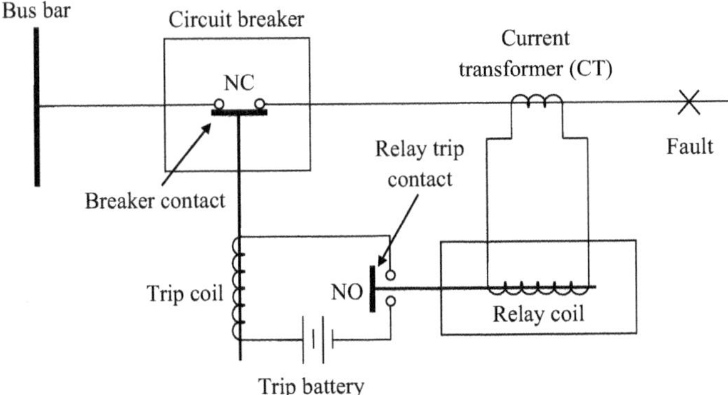

FIGURE 9.31 Simplified relay circuit [III].

circuit breaker and opening of normally closed circuit-breaker contacts. The circuit breaker then opens and isolates the faulted portion or component from the rest of the power system. The relay thus protects the transmission line from damage occurring due to fault and it also permits proper functioning of healthy portion of the power system.

Each of the three main components of power system protection mentioned above is briefly described below.

9.8.1 Instrument Transformers (Transducers)

A current or voltage transformer is a device which reproduces in its secondary winding a current or voltage which is proportional to primary current or voltage. In a protection system, current or voltage transformers convert measured currents and voltages of the protected power system equipment to low level proportional signals for relay operation. The reduction in the measured system voltages and currents to low level signals is necessary since low level current and voltage input to the relays leads to requirement of small and less expensive physical hardware for relay construction, and personnel working with the relay also have safe work environment [22,38].

A schematic illustrating arrangements of a current transformer (CT) and a voltage transformer (VT) in a simple power system circuit is shown in Figure 9.32.

The CT reduces the primary current to low level proportional current signal and the VT reduces the primary voltage to low level proportional voltage signal for relay operation. The standard current rating of CT secondary in the USA is 5 [A], whereas the standard VT secondary voltage rating is 115 [V] (line-to-line). The

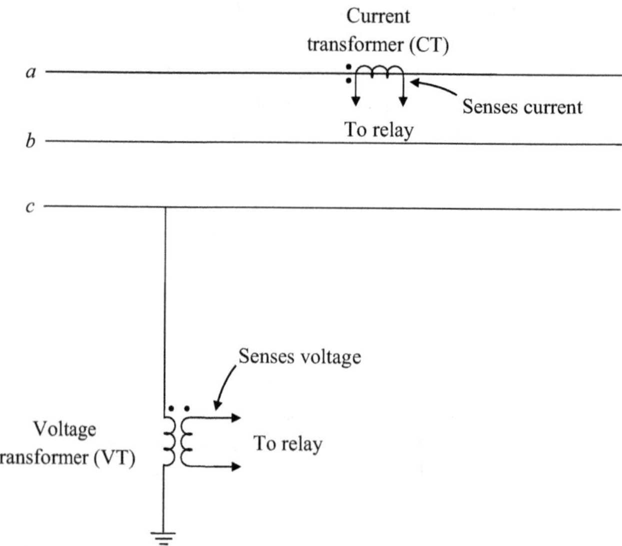

FIGURE 9.32 Arrangement of CT and VT in a simple power system circuit.

standard CT ratios available are 50:5, 100:5, 150:5, 200:5, 250:5, 300:5, 400:5, 450:5, 500:5, 600:5, 800:5, 900:5, 1000:5, 1200:5, 1500:5, 1600:5, 2000:5, 2400:5, 2500:5, 3000:5, 3200:5, 4000:5, 5000:5 and 6000:5. The standard VT ratios available are 1:1, 2:1, 2.5:1, 4:1, 5:1, 20:1, 40:1, 60:1, 100:1, 200:1, 300:1, 400:1, 600:1, 800:1, 1000:1, 2000:1, 3000:1 and 4500:1 [16].

9.8.2 Relays

Relays are devices which process the voltage and/or current signals received from the instrument transformers to detect the fault and permit the circuit-breaker trip circuit to be energized, which results in opening of circuit-breaker contacts [16,38]. In order to perform the function properly, a relay must have the main characteristics of sensitivity, selectivity and speed. Sensitivity means that the relay must be able to operate when a fault results in minimum fault current flow through the relay. The selectivity of a relay is its ability to recognize and locate the fault. Relays must be able to distinguish between the faults in their own protected equipment for which they must trip and faults in adjacent equipment for which they should not trip. Some relays are inherently selective, whereas others relays are not. A relay is also required to operate with proper speed. Speed is necessary in isolating the faulted equipment or section of the power system, since it has direct relationship with the damage occurring due to the fault, and consequently the cost and delay in carrying out the repairs.

Some of the main types of relays include over-current relays, directional relays, distance or impedance relays, differential relays, pilot relays and digital or microprocessor-based relays.

Over-current relays receive input signals from current transformers. They operate when the secondary current value in the CT supplying current to the relay is more than relay pickup setting. The relay can then send instantaneous trip signal to the circuit breaker to disconnect the faulted equipment or section as early as possible. There can also be intentional time delay introduced in sending the trip signal for the purpose of achieving relay coordination. To understand the requirement for relay coordination, we refer to Figure 9.33. In the simple power system shown, the generator supplies power to loads A, B and C. There are four circuit breakers at locations 1, 2, 3 and 4. Each of these circuit breakers has separate CT and over-current relay. When a fault occurs at *F*, circuit breaker 2 should operate as early as possible and disconnect the faulted section. The other circuit breakers in the power system should

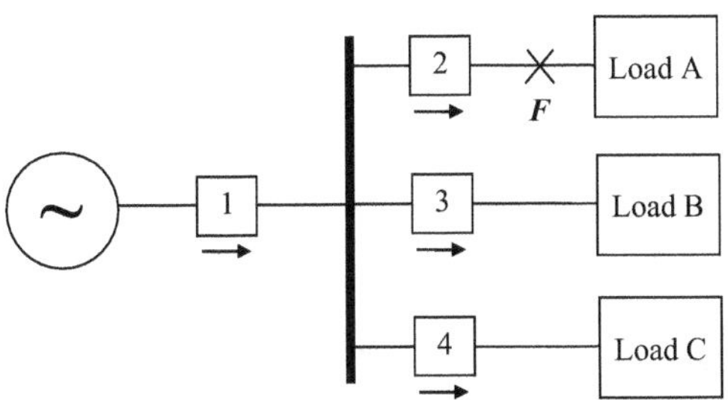

FIGURE 9.33 Time-delay over-current protection [38].

remain closed. Load A will therefore become disconnected due to fault, however loads B and C will continue to work.

For the fault condition shown, circuit breakers 3 and 4 will not operate since their relays will not sense the fault. Relays connected to circuit breakers 1 and 2 will both sense the fault condition and currents in them will be more than their pick up settings. It is desirable that circuit breaker 2 should operate earlier than circuit breaker 1 to clear the fault. It will prevent unnecessary disconnection of loads B and C. Tripping of circuit breaker 2 prior to circuit breaker 1 is achieved by introducing intentional time delay in the operation of relay associated with circuit breaker 1. The time delays introduced in relay operation however must be properly coordinated with fault current magnitudes since larger faults must be cleared as early as possible to minimize the damage occurring due to the fault.

A directional relay can detect fault current flowing in one direction only. It will operate to isolate the faulted section or equipment located on its left side or right side if fault current flow is in the specified direction. Its operation is determined from phase angle of the fault current with respect to the reference voltage. A 180° range of phase angles is defined for directional relay operation with respect to the reference voltage. If fault current has phase angle within the range specified, the relay will operate and trip the circuit breaker. If phase angle of the fault current is not within the range specified, the relay will not operate and trip signal will not be sent to the circuit breaker.

The operation of a distance or impedance relay depends upon the ratio of voltage and current at the relay location. Impedance is the ratio of voltage and current, and the amount of impedance between relay location and fault is dependent upon the length of transmission line between these two locations. This type of relay can be configured to operate for faults occurring within a set distance from the relay and trip the circuit breaker. For faults occurring beyond the set distance, distance relay will not operate and trip signal will not be sent to the circuit breaker. Distance relays can be installed at other locations to protect other portions of the transmission line, and provide relay coordination and backup.

The protection of transformer using differential relay is shown in Figure 9.34. The CT turns ratio on the primary and secondary side of the transformer are chosen such

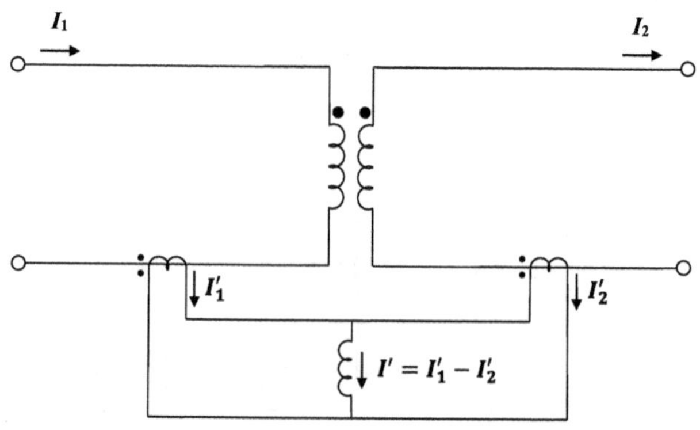

FIGURE 9.34 Differential protection of a transformer [38].

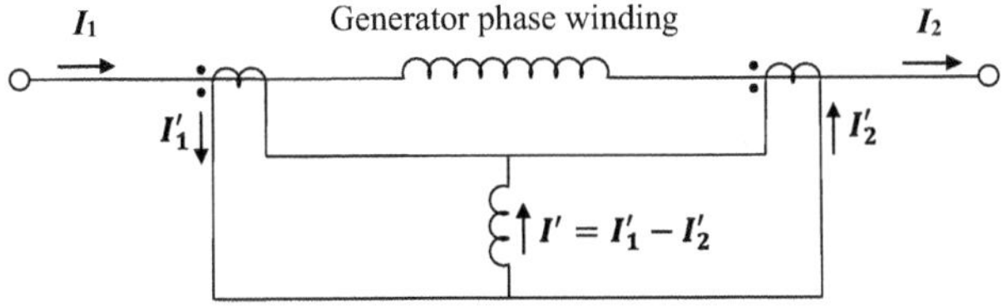

FIGURE 9.35 Differential protection of a generator [38].

that for proper operation of the transformer, the current flowing through differential relay coil $I' = I'_1 - I'_2 = 0$. During transformer fault, $I' = I'_1 - I'_2 \neq 0$ and current flowing through the differential relay coil has non-zero value. If I' value is greater than differential relay pickup current setting, it will operate to disconnect the circuit breaker and isolate the faulted equipment. The differential protection scheme for a generator is shown in Figure 9.35. The CTs on the two sides of the generator phase winding have the same turns ratio.

Pilot relaying is also a differential protection technique that, as opposed to the above-described direct wire connection of the relays, compares the quantities at the terminals using a communication channel. Since the devices such as generators and transformers are installed in a single location where it is possible to directly connect CTs and relays, pilot relaying is not required for differential protection of these devices. The transmission line terminals are located many kilometers apart, and therefore for differential relaying protection of transmission lines, pilot relaying is required. For fault detection in transmission lines using pilot relaying, two most common methods are directional comparison and phase angle comparison. In directional comparison, power flows at the two terminals of the transmission line are compared for fault detection. In phase angle comparison, current phase angles at the two terminals of the transmission line are compared for fault detection.

Digital relays are electronic devices which receive voltage and/ or current signals from instrument transformers and process them in some way to decide fault condition or to provide relay coordination. For instance, they can calculate negative- and zero-sequence components of currents to determine fault condition and isolate the faulted section from rest of the power system. A main benefit of digital relaying is that by updating the relay software, it is possible to alter relay characteristics such as conventional relay settings and block/trip regions [16,38].

9.8.3 Circuit Breakers

A circuit breaker is a device which performs the function of disconnecting the faulted equipment or section from the power system within few cycles after receiving the trip signal from the relay. The circuit breaker has the ability to perform the intended function of clearing the fault without damage to itself.

During the opening of current carrying contacts in a circuit breaker for fault clearance, the high electric field between the contacts results in ionization of gas between them. The severity of the arcing increases with increase in voltage difference between the parting contacts. The following two main principles are utilized

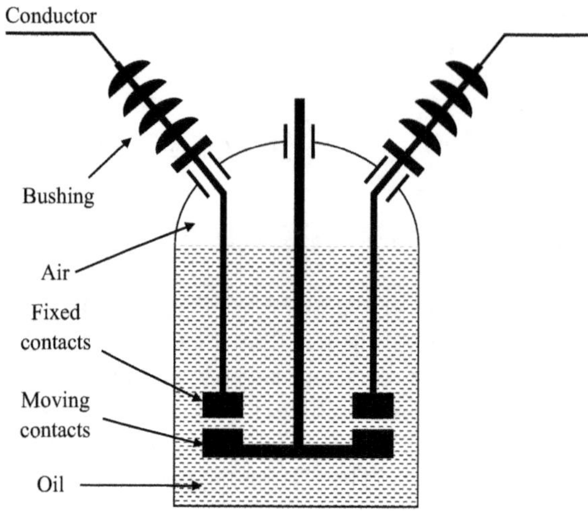

FIGURE 9.36 Components of a typical oil circuit breaker [IV].

by all circuit breakers for quenching the arc. The first principle involves lengthening the arc until it becomes long and thin, which results in increase in arc resistance and decrease in arc current and temperature. Eventually, there isn't sufficient energy in the arc to keep it ionized. The second principle involves opening the contacts in an energy absorbing medium which results in cooling and quenching of the arc. As the arc quenching medium, air, oil, and insulating gas (such as SF_6) are commonly employed.

Figure 9.36 shows the components of a typical oil circuit breaker. Its control by a relay circuit is shown in Figure 9.31 earlier. The main components of the oil circuit breaker are movable and fixed contacts enclosed in a strong metallic tank and submerged in oil.

During proper power system operating conditions, the circuit-breaker contacts remain closed and the full-load current flows through the circuit breaker continuously. The voltage which is induced in the secondary winding of the current transformer of relaying circuitry is inadequate in these conditions to energize the circuit breaker's trip coil. During fault, high current in current transformer primary winding induces high voltage in current transformer secondary winding. As a result, the relay coil is energized which then energizes the trip coil of the circuit breaker as described earlier. The energized circuit-breaker trip coil opens the circuit-breaker contacts thus disconnecting the faulted portion or equipment from the rest of the power system. The circuit breaker oil quenches the arc produced during the opening of contacts.

9.9 FUSES

A fuse is simply a small length of wire or thin strip which melts when excessive current flows through it for sufficient time during short-circuits and sustained overloads. It is installed in series with the AC or DC circuit to be protected. During proper operating conditions of the power system, the temperature of the fuse element is below its melting temperature value, and load current flows through the fuse without overheating it. During faults and sustained overloads, the current flowing through the fuse is more than its rated capacity. This results in increase in fuse temperature beyond its melting temperature value and the fuse melts to disconnect the circuit it is installed to protect [22,35,38]. The response of fuse is such that the higher the current value, the lesser the time required for circuit opening, which represents an inverse-time characteristic. The fuse thus protects power system equipment from damage due to flow of excessive currents during faults and sustained overloads. It should be noted that a fuse performs both detection and interruption functions.

9.10 PROTECTION OF GENERATORS, TRANSFORMERS, BUSBARS AND TRANSMISSION LINES

Differential relays described earlier provide protection against some types of faults to transformers and generators [38]. The differential relay arrangements for fault protection in a transformer and generator were shown in the Figures 9.34 and 9.35 respectively. The similarity between the protection arrangements for transformer and generator shown should be noted. There is direct relationship between fault current and differential current $\boldsymbol{I}' = \boldsymbol{I}_1' - \boldsymbol{I}_2'$ flowing in the differential relay coil. If $\boldsymbol{I}'$ value is greater than differential relay pickup current setting, it will operate to disconnect the circuit breaker and isolate the faulted equipment in case of an internal fault. External faults have no effect on the operation of differential relays and there relays do not trip in case of an external fault.

Some other relays used for generator protection include relays for protection against over-speeding, protection against over-voltages and protection against unbalanced loading when there are different phase currents in the stator windings. Some other relays used for transformer protection include relays for protection against overloads and protection against incipient or slow-developing faults.

Differential relaying scheme is also utilized for protection of busbars as shown in Figure 9.37. In this scheme, current entering and coming out of the bus are summed. During proper operation of the busbar, the current flowing through differential relay coil $\boldsymbol{I}' = \boldsymbol{I}_1' - (\boldsymbol{I}_2' + \boldsymbol{I}_3') = 0$. During busbar fault, $\boldsymbol{I}' = \boldsymbol{I}_1' - (\boldsymbol{I}_2' + \boldsymbol{I}_3') \neq 0$ and current flowing through the differential relay coil has non-zero value. If $\boldsymbol{I}'$ value is greater than differential relay pickup current setting, it will operate to disconnect the circuit breaker and isolate the faulted busbar.

The transmission line is generally very long and it runs through open atmosphere. The probability of occurring fault in a transmission line is much higher than that in a generator and a transformer. Consequently, transmission lines require much protection. Protection of transmission lines should have some special features too. During fault, the only circuit breaker closest to the fault location should trip. If the circuit breaker closest to the fault fails to trip, the circuit breaker just next to this breaker should trip as back up. The operating time of relay associated with protection of transmission line should be as less as possible in order to maintain power system stability and prevent unnecessary tripping of circuit breakers associated with other healthy portions of power system.

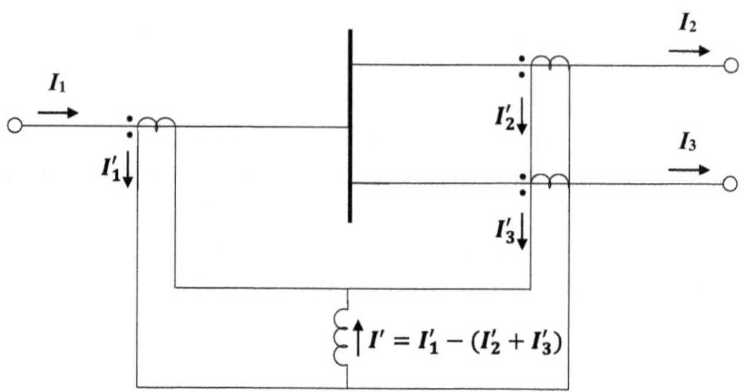

FIGURE 9.37 Differential protection of a busbar [V].

The main relaying schemes used for transmission line protection include time-delay over-current relaying, directional over-current relaying, impedance relaying and pilot relaying.

Figure 9.38 demonstrates the protection of radial transmission line with time-delay over-current relays. The relays provide primary protection to one line and they also provide remote backup protection to a neighboring line. The direction of protection of relays is indicated using arrows. For instance, relay at bus 1 will provide primary protection to transmission line between buses 1 and 2, and it will provide backup protection to transmission line between buses 2 and 3. Also, for a fault between buses 2 and 3, relay at bus 2 should operate first and there should be adequate time delay in the operation of relay at bus 1 to maintain relay coordination [37,38].

Directional over-current relays with coordinated time-delay settings are utilized for the protection of transmission lines fed from both ends, parallel transmission lines and loop systems. To make an over-current relay at a location directional, a directional relay is also provided at that location. The outputs of directional relay and over-current relay are arranged such that circuit breaker will operate only when both relays provide the trip signal. For the protection of transmission lines in a complex interconnected system, impedance relays are many times utilized which are configured to operate for faults occurring within a set distance from the relay.

The student reader is requested to refer to any standard power system protection reference for detailed description about power system protection.

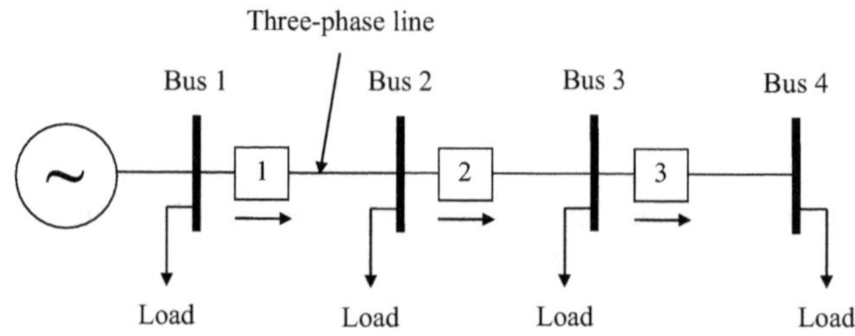

FIGURE 9.38 Radial transmission system protection using time-delay over-current relays [38].

9.11 POWER SYSTEM STABILITY

Power system stability refers to the ability of the power system to maintain synchronism between its generators and remain in an operating equilibrium despite the disturbances inevitable in any network consisting of power plants supplying electrical energy to the loads [9,16,38]. The disturbances can be small such as those occurring during variations in load on the power system or they can be large such as those occurring during power system faults. A stable power system returns to a state of acceptable voltages and power flows throughout the system after the disturbance.

There are three types of power system stability: steady state, transient, and dynamic.

Steady-state stability deals with the response of synchronous generators in a power system to power system load variations and more precisely to relatively small power system disturbances. It is known as steady-state stability since generators are assumed to maintain synchronism and supply power to loads that are much below their maximum capabilities and the disturbances occurring will not stress the machines to their maximum capabilities.

In steady-state stability, operating variables vary gradually or slowly. Steady-state stability studies are carried out mainly with the help of a power-flow computer program, and they ensure proper power system operating conditions such as nominal values of power system bus voltages, not too large phase angles across the transmission lines, and non-overloaded operation of generators, transformers, transmission lines, and other power system equipment.

Transient stability deals with maintaining synchronism among all synchronous generators in the power system during occurrences of severe disturbances such as faults due to lightning strikes, abrupt removal of a generator or a transmission line from the power system network, or any severe shock to the power system resulting from a switching operation. Due to the severity and abruptness of the disturbance, transient stability study deals with power system operation during the first few seconds after the occurrence of fault or switching activity. Due to the short duration of transient stability study, synchronous generator deviations from synchronism are assumed to be small and all voltages, currents, and power system parameters considered in the analysis are known as 60-Hz values.

Due to power system disturbances, synchronous generators experience transient deviations from synchronous frequency, and their power angles vary. Transient stability analysis is carried out to determine whether or not synchronous generators will resume synchronized operation at synchronous frequency with new values of steady-state power angles [9,16,38]. Such analysis is mainly performed with the help of digital computers since real power systems are extremely large with many synchronous generators and numerous transmission lines.

Dynamic stability analysis is carried out for an even longer time duration of several minutes. Even if transient stability is maintained, the transients that occur in power system during transient stability can be present for a long time up to many minutes. The effects of regulating systems such as automatic frequency-control loops and automatic voltage-control loops must be considered during this prolonged duration. The result is a more complex network configuration with altered time constants which result in alteration of the associated transient behavior of the power system [38]. Dynamic stability analysis deals with the behavior of the power system under these circumstances.

The reader is requested to refer to standard power system references mentioned in the list of references for a detailed description about power system stability.

9.12 SUMMARY

Different kinds of faults which occur in a power system are broadly classified into symmetrical faults and unsymmetrical faults. Symmetrical faults such as three-phase faults are generally analyzed on a per-phase basis. In case of unsymmetrical faults such as line-to-ground and line-to-line faults, the method of symmetrical components is utilized to determine voltages and currents in all parts of a power system during unbalanced system conditions.

The purpose of a power system protection is to sense the fault condition and disconnect the faulted equipment or section as quickly and as selectively as possible, thus preventing or minimizing harmful effects of faults. Many protective relays operate from the symmetrical component quantities. A good understanding of symmetrical components is therefore extremely useful in the study of power system protection.

The ability of synchronous generators to move from one steady-state operating state after the occurrence of a disturbance to another steady-state operating state, without loss of synchronism is called power system stability. Power system stability is of three types: steady-state stability, transient stability, and dynamic stability.

For a more detailed description of power system fault analysis, power system protection, and power system stability, the reader is requested to refer to power system references mentioned in the list of references.

REVIEW QUESTIONS

9.1 What is a power system fault?

9.2 What are symmetrical power system faults?

9.3 What is the difference between a symmetrical fault and an unsymmetrical fault?

9.4 What are the different types of unsymmetrical faults occurring on the power systems?

9.5 Describe the procedure for symmetrical fault analysis of electrical power systems?

9.6 Describe the purpose and usefulness of symmetrical components in fault analysis?

9.7 What are sequence networks?

9.8 Describe the procedure to determine the fault current in case of a single line-to-ground fault on an unloaded generator.

9.9 Describe the procedure to determine the fault current in case of a line-to-line fault on an unloaded generator.

9.10 Describe the procedure to determine the fault current in case of a double line-to-ground fault on an unloaded generator.

9.11 Describe how sequence networks are connected to determine the fault current in case of a single line-to-ground fault on a power system.
9.12 Describe how sequence networks are connected to determine the fault current in case of a line-to-line fault on a power system.
9.13 Describe how sequence networks are connected to determine the fault current in case of a double line-to-ground fault on a power system.
9.14 What is the purpose of power system protection?
9.15 What are the main requirements of a properly designed protection system?
9.16 What are the main components of a protection system?
9.17 Draw a typical relay circuit for over-current protection and describe its working.
9.18 What are different types of instrument transformers and what is their purpose in a protection system?
9.19 What are the standard voltage transformer and current transformer ratios available?
9.20 Describe what is relay and what its function is in a protection system.
9.21 Describe the working principle of a directional relay.
9.22 Describe the working principle of a distance relay.
9.23 Describe the working principle of a differential relay.
9.24 Write a short note on pilot relaying.
9.25 What is the difference between a fuse and a relay?
9.26 Describe the purpose of a circuit breaker in a protection system and its working principle.
9.27 Describe the relaying schemes available for the protection of generators.
9.28 Describe the relaying schemes available for the protection of transformers.
9.29 Describe the differential relaying scheme used for the protection of busbars.
9.30 Describe the relaying schemes available for the protection of transmission lines.
9.31 What is meant by power system stability?
9.32 Describe the different types of power system stability.

10 Future of Electrical Energy

10.1 INTRODUCTION

Conventional fossil fuels, such as coal, natural gas, and oil, which have been used by humanity to power societies since the industrial revolution, have several problems associated with them. Their resources are also extremely limited and depleting rapidly.

Burning of fossil fuels releases particulate air pollutants and greenhouse gases into the atmosphere. Particulate air pollutants pollute the cities with smog. Greenhouse gases are primarily responsible for global warming. Combustion of fossil fuel also results in release of significant amount of heat into the environment, mainly due to the inefficiency of the energy conversion process.

Therefore, work is in progress to replace fossil fuels with energy sources of the future for the generation of electrical energy. The electrical energy of the future will be generated from renewable sources. Solar energy, wind energy, wave and tidal energy, geothermal energy, hydroelectric power, nuclear power, and energy from biomass and biofuels will be utilized for generation of electrical energy of tomorrow. The internal combustion engines and other gas-powered motors will be discontinued along with fossil fuels they require, and they will be replaced with electrically powered machines.

This chapter describes the environmental effects of present electrical power generation, transmission and distribution systems, and various renewable electrical energy sources of the future. Economics of renewable energy systems and electrification of industrial processes and transportation facilities are also briefly described.

10.2 ENVIRONMENTAL EFFECTS OF ELECTRICAL ENERGY GENERATION

At the place of utilization, electrical energy is pollution free and relatively safe form of energy. The generation, transmission, and distribution of electrical energy, however, have adverse environmental impacts [VIII]. The effects of different types of power plants and associated transmission and distribution systems on the surrounding environment are described below.

10.2.1 Effect on Landscape

Power plants are generally large structures located on hundreds of acres of land area, and they result in the alteration of visual landscape. The impact of a power plant on the surrounding landscape increases with the plant size. To build a power plant,

DOI: 10.1201/9781003432340-10

clearing of the land area is necessary. Besides access roads and electricity transmission lines, some power plants also require arrangements for fuel supply and storage and cooling water supplies. An area to store the ash produced after fuel combustion is needed for power plants, which burn solid fuels, such as coal.

Power transmission lines and distribution systems that transmit electricity from power plants to consumers also have adverse effects on the surrounding landscape. The transmission lines are supported above ground using large transmission towers. Alteration of the visual landscape occurs due to transmission towers and power transmission lines particularly when they cross undeveloped areas. Locating power lines below ground is possible; however, doing so is more expensive and is rarely practiced outside of urban areas.

10.2.2 Air Pollution

Globally, almost two-third of the total electricity generation in 2021 was from fossil fuels, such as coal, natural gas, and petroleum. Carbon dioxide (CO_2), carbon monoxide (CO), sulfur dioxide (SO_2), nitrogen oxides (NO_x), particulate matter (PM), and heavy metals like mercury and lead are the combustion byproducts present in combustion gases that are produced when these fuels are burned.

Almost all combustion byproducts are harmful to the environment and health of mankind. Carbon dioxide released during combustion results in greenhouse effect and causes global warming. CO is a colorless and odorless gas, and it is harmful when inhaled since it displaces oxygen in the blood and deprives the heart, brain, and other vital organs of oxygen. Inhaling large amount of CO can cause loss of consciousness and suffocation in minutes. Acid rain is caused due to SO_2 emission, and it is harmful to plants and aquatic animals. Worsening of respiratory illnesses and heart diseases also occurs due to inhalation of SO_2. PM adversely affects visibility in cities and scenic areas and also causes respiratory diseases. Heavy metals, such as mercury and lead, found in combustion byproducts are extremely harmful for humans and animals.

The amount of some of the aforementioned compounds that power plants can emit into the atmospheric air is restricted by air pollution emission standards. Some of the methods which are practiced to maintain compliance with the standards include burning of low-sulfur-content and pre-treated or processed coal to reduce SO_2 emissions, the installation of various kinds of particulate emission control devices to treat combustion gases prior to exit from the power plant, the use of wet and dry scrubbers containing lime to reduce SO_2 emissions, and the use of selective catalytic and non-catalytic converters to reduce NO_x emissions [VIII].

10.2.3 Solid and Liquid Wastes

After combustion of solid fuels, such as coal, ash is the solid waste that is left behind. The largest particles accumulating at the bottom of the combustion chamber of power plant boilers result in bottom ash. The lighter and smaller particles accumulating in air emission control equipment result in fly ash. It is a common practice to mix bottom ash with fly ash. The collected ash contains several hazardous substances, and it is piled up in large heaps close to the thermal power plants. Coal-fired power plants sometimes sell the collected ash for the utilization in landfills and for the production of concrete blocks.

10.2.4 Nuclear Wastes

Electrical power generation in nuclear power plants does not result in emission of greenhouse gases, PM, SO_2, or NO_x. However, nuclear power plants produce low-level radioactive wastes and high-level radioactive wastes. Low-level wastes generated during the working of a nuclear power plant, such as contaminated protective clothing or machine parts exposed to radioactivity, are stored at the plant to allow radioactivity to decay to safe levels. Depending on the residual radioactivity level, they are then either disposed as ordinary trash or sent to a low-level radioactive waste disposal site. High-level wastes consist of highly radioactive spent nuclear fuel assemblies. They are stored in specially designed storage containers and facilities and subsequently disposed of in a way that provides adequate protection to mankind.

10.2.5 Noise Pollution

Regular exposure to noise emanating from power plants from the use of equipment, such as boilers, turbines, and coal crushers, can cause noise-induced hearing loss and other adverse health effects to people working in the power plants and residing in communities located close to the power plants.

10.3 RENEWABLE ENERGY

Since solar energy will be available as long as the Sun shines, it is referred to as renewable energy. The other main renewable energy sources are wind, hydro, waves and tides, geothermal, and biomass and biofuels. These renewable energy sources are briefly described below. Even though nuclear energy is not renewable, it is also described since the amount of uranium present on the Earth is enough to provide energy for billions of years.

10.4 RENEWABLE ENERGY SOURCES OF THE FUTURE

10.4.1 Solar Energy

Solar power is the most promising renewable energy source. Enough energy falls on the Earth's surface in the form of sunlight in a single hour to power all of present civilization for almost one year. Solar power systems accounted for around 4% of the world's electrical energy generation in the year 2021, and their contribution is increasing rapidly. Solar energy can be converted into electrical energy using the following two different approaches: concentrating solar power technologies and photovoltaic (PV) conversion.

In concentrating solar power technologies, the Sun's rays are concentrated by lenses or mirrors on to a receiver containing heat transfer fluid, which is heated and then used to raise steam for driving steam turbines. The steam turbines provide motive power to the three-phase synchronous generators to generate three-phase electrical power. To concentrate the Sun's rays, the lenses and mirrors used must have accurately curved surfaces, and they also require tracking system to follow the motion of the Sun. The heat transfer fluid can also be used to store energy from the sunlight, allowing power generation at night and on cloudy days.

In PV conversion, solar energy is directly converted into electrical energy in PV cells using semiconductors. Electrons are released when sunlight hits the semiconductor in the PV cell, and then, they result in an electric current. PV cells have no moving parts, and there are no thermal stresses on the system. The PV cells work quietly, and they are extremely reliable with long lifespan and require very less maintenance.

Individual solar PV cells are the building blocks of the PV system. An individual solar cell has square cross section and approximately 6 inches in size. The voltage output of a solar cell is around 0.5–0.6 [V], and the power output is around 1–2 [W], which is not sufficient to do any useful work. Therefore, solar cells are connected together and mounted on a rectangular frame to produce PV panel, and PV panels are connected together to form PV module for higher voltage and power outputs [XII]. Several modules are then wired together to produce an array which can be scaled up or down to provide the required amount of power at a proper voltage level as shown in Figure 10.1.

Solar cells can be fabricated from several semiconductor materials. Silicon is the most common semiconductor material used in the fabrication of solar cells; however, other semiconducting materials, such as cadmium telluride (CdTe) and copper indium gallium diselenide (CIGS), are also being tested to improve the efficiency of sunlight to electricity conversion process. The efficiency at which PV cells convert

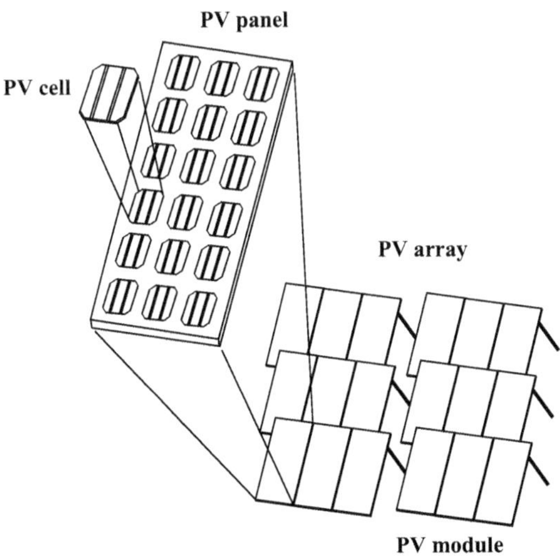

FIGURE 10.1 Photovoltaic system [XIV].

sunlight into electricity varies depending on the semiconductor material, and it ranges between 15% and 30% in present time.

PV cells produce direct current (DC) electricity. It can be used to charge batteries, which subsequently supply power to devices using direct current electricity. Nearly, all electricity is supplied as alternating current (AC) in power transmission and distribution systems as described in previous chapters. The DC electricity is converted to AC electricity using inverters to supply power to devices using AC electricity.

PV cells facing the Sun directly generate the maximum amount of electricity. Tracking systems can be used with PV modules to rotate them, so they are always facing the Sun; however, such tracking systems are costly. PV systems generally have modules in fixed position facing true geographical south in the northern hemisphere and true geographical north in the southern hemisphere at a tilt angle nearly equal to the geographical latitude at that location since it maximizes their performance and results in optimal annual electrical energy production [XII,XIII].

The number of houses powered by housetop-mounted solar panels and connected to the local electricity grid has substantially increased in recent years. Solar panels are linked to the local electrical grid network in a grid-connected PV system through high-quality DC to AC inverter and filter system, enabling them to function alongside the grid. During the daytime, grid-connected PV systems provide some or all of the power needed for houses. At night, however, when solar panels do not produce any electricity, houses receive power from the local electrical grid network. When the Sun is shining brightly during a long hot summer day, PV systems can generate more electricity than is required in houses. This extra electricity is supplied to the local electrical grid network. Grid-connected PV systems, therefore, receive power from the grid or supply power to the grid depending on the sunlight conditions and the actual electrical demand. Connection of the solar panels to the local electrical power grid also provides net billing facility. If the solar PV system produces more electricity on a hot summer day than is required, the extra electricity is fed back to the utility grid, which causes the electricity metering system in the house to spin backward. During such an occurrence, the local utility company provides credit to the customer for the electricity supplied to the grid. Consumers are billed for the net electricity consumed if they consume more electricity than is generated in their grid-connected PV system [XI].

10.4.2 Wind Energy

Wind power generation systems utilize wind turbine connected to generator and energy of the wind to produce electricity. Wind power is a popular, sustainable, renewable energy source that has negligible impact on the environment in comparison to burning fossil fuels. Wind energy systems contributed to around 6.5% of the world's power demand in the year 2021. New wind turbine power generation systems are rapidly being installed around the world. Global installed wind-generation capacity has increased from 7.5 [GW] in 1997 to around 750 [GW] in 2021. There are powerful winds in many parts of the world; however, the best locations for wind turbine power generation are sometimes in remote areas and offshore areas.

The kinetic energy of the wind is converted into electrical energy using wind turbines or wind energy conversion systems. When wind hits the turbine's blades, it causes them to rotate and the turbine connected to them. The turbine in turn spins a generator that generates electricity. Simple and small battery-charging wind turbines use DC generators; however, more common grid-connected wind turbine power systems utilize AC generators.

The size of the wind turbine and the length of its blades determine how much electricity can be obtained from the wind. The output is proportional to the rotor size and cube of the wind speed. Doubling the wind speed theoretically results in eight times increase in power output.

Large wind turbine systems almost always have horizontal axis with upwind rotors consisting of three blades mounted on a tall tubular tower. The average rotor diameter in present time is around 125 [m], average hub height is around 90 [m], and wind turbines typically rotate at 15–20 [rpm]. Turbine capacities for wind power projects in present times are about 3 [MW] onshore and 10 [MW] offshore. Wind turbine systems have an efficiency of around 30%–45%.

Low wind speeds do not produce any electrical power since there is insufficient energy to power the turbine. The power generated tends to increase as the cube of wind speed when velocity rises above the cut-in wind speed, which is between 3 and 5 [m/s]. The generator is generating the rated power when winds attain the rated wind speed. The turbine is controlled, usually by changing the blade pitch angle, at higher wind speeds to produce the rated output up to a maximum wind speed. The wind turbine system is shut down after the wind speed reaches cut-out wind speed or furling wind speed in order to prevent damage from excessive wind loading [31].

The wind turbine rotates at speeds that will generate the maximum possible amount of power. The frequency of the electrical output from the AC generator connected to the wind turbine will, however, vary depending on the current speed of the wind turbine. Therefore, it is not possible to connect the generator directly to the grid which in the USA requires fixed 60 [Hz] frequency. To overcome this problem, high-power transistors initially rectify variable frequency AC power from the generator and convert it into DC power. Using an inverter, this DC power is then converted back into AC power at a constant 60 [Hz] frequency. The somewhat uneven output of the inverter is filtered to create a smooth AC waveform [31].

A wind farm has many wind turbines in an area. A large wind farm can have several hundred individual wind turbines distributed over a large area. The land area between the turbines can be used for agricultural or other purposes. A wind farm can also be located offshore.

10.4.3 Hydroelectric Energy

Hydropower is generated by hydroelectric power plants, in which river water powers large hydroturbines which in turn drive three-phase synchronous generators to produce electrical power in three-phase form. The greater the water flow and the higher the head of water, the more is the electricity which can be generated. Since hydroelectric power uses water to generate electricity, plants are usually located close to sources of water. Hydroelectric power plants have been described earlier in

Chapter 2. Hydroelectric power plants currently supply around 16% of the world's electrical energy requirement, and they will continue to do so in the future.

10.4.4 Nuclear Energy

Nuclear power plants utilize uranium fuel and the process of nuclear fission for electrical energy generation. The splitting of uranium atoms during the process of nuclear fission releases a large amount of energy in the form of heat and radiation. The heat produced during nuclear fission is used to heat water, and the steam produced is used to spin large steam turbines, which in turn drive three-phase synchronous generators to produce three-phase electrical power. Nuclear power plants have been described earlier in Chapter 2.

Despite the dangers involved in nuclear power generation, 10% of the world's electricity demands are met from nuclear power plants, and this trend will continue. Nuclear power will have significant contribution in the generation of electrical power free from carbon dioxide emissions. More nuclear power plants will be constructed in the future; however, not many since safer and truly renewable alternatives like wind and solar energy systems will be utilized more commonly for electrical energy generation.

10.4.5 Geothermal Energy

Geothermal energy is a renewable source, which utilizes the enormous heat coming from the Earth's crust. In some areas of the world, hot springs or geysers and molten lava streams are released close to the ground surface, which can be used for steam production for generating electricity. Due to the lower steam pressure and temperatures, the efficiency is less than thermal power plants; however, the capital costs are less, and the fuel is certainly free. The installed electrical power generation capacity of geothermal energy sources worldwide in the year 2021 was around 16 [GW], and this capacity will gradually increase in future.

10.4.6 Wave and Tidal Energy

Waves are created when wind blows over the surface of open ocean water. The wave motion can be converted into mechanical energy in several ways with a number of innovative solutions being tested. The wave energy can be utilized by allowing it to flow into a narrow channel to increase its size and power, which can then rotate the turbines for electricity generation. It is also possible to channel waves into a catch basin or reservoir from where the water is allowed to flow into a turbine located at a lower elevation to generate electrical energy similar to hydroelectric power plants [XV]. Every wave power system must have the ability to withstand the tremendous loads that occur during storms.

Another renewable source is tidal power. At high tide, seawater is allowed to flow into a basin, and it operates the turbines for electrical power generation. At low tide, the stored water is released, operating the turbines once more.

The installed electrical power generation capacity of wave and tidal energy sources worldwide in the year 2021 was less than 1 [GW], and this capacity will increase in future.

10.4.7 Energy from Biomass and Biofuels

Biomass is the vegetable matter produced in forestry and agriculture. It has carbon-based organic composition, and during combustion, it combines with oxygen to release heat. If the heat released results in temperature levels above 400 [°C], it can be used in a furnace to produce steam that will power steam turbines to generate electrical energy. A biofuel is fuel, mainly a liquid or gaseous fuel, produced from biomass after transformation using chemical and biological processes. Methane gas and liquid ethanol are examples of biofuels produced from biomass. Biofuels can also be used for the generation of electrical energy in power plants. The energy generated from biomass and biofuels is collectively called bioenergy [47]. Bioenergy contributes to around 2% of world's electrical power demand at present, and this percentage will increase in the future.

10.4.8 Electrical Energy from Renewable Sources

The electrical energy in future will, therefore, be generated from a mix of renewable energy sources described above, which currently provide little more than one-third of the world's current electrical power requirement [IX]. Future improvement in the efficiency of electrical appliances will also help in reducing energy requirement.

10.5 RENEWABLE ENERGY ECONOMICS

When determining whether it is financially beneficial to install renewable energy systems, factors, such as initial installation costs and net annual electrical energy production, are considered. Electrical energy from a renewable source must compete with electrical energy generated from competing technologies when determining its economic feasibility [39,42]. Renewable energy systems must generate enough returns from the electricity produced, which must exceed all costs in a reasonable duration of time for economic viability.

The value of electricity generated from renewable energy systems depends on the resource utilized. To determine the renewable energy system that generates the required electrical energy at the lowest cost among the different alternatives, economic analysis is conducted. Various possible variants of renewable energy systems are, therefore, compared during the estimation of costs of electricity generation. It is also a common practice to compare renewable energy systems with conventional electrical energy systems, although many of these comparisons do not consider all advantages connected with renewable energy systems, such as environmental pollution control and improved health in societies [39,42]. Renewable energy economics is described in detail in standard textbooks on renewable energy systems.

10.6 ELECTRIFICATION

Electricity has been used historically to power-limited number of processes and systems. Fossil fuel engines are used in a number of industrial processes and transportation systems. With advancement of technology, however, electricity driven machines can gradually replace fossil fuel engines for powering these industrial processes and transportation systems [X]. This practice is called electrification.

The car is a good example of electrification. Until few years earlier, internal combustion engines were used to power all cars. Advancements in battery energy storage systems have resulted in increasing popularity of electric vehicles, and they are gradually replacing internal combustion engine powered vehicles. With time, electric vehicles will completely replace internal combustion engine powered vehicles, and the demand for electricity will rise during this conversion process.

Due to recent technological advancements, it has increasingly become possible to power more processes and systems using electricity that previously required fossil fuel powered engines. As a result of further electrification, there will be a less requirement of burning fossil fuels locally, and electricity will be utilized instead. Electrification, if done properly, will lead to a large decrease in world carbon emissions [X]. It is possible to further reduce carbon emissions in all parts of the economy if developed countries stop burning fossil fuels and rely on renewable energy sources for electrical energy generation to power various processes and systems.

10.7 SUMMARY

Fossil fuels like coal, oil, and natural gas supply nearly two-third of the world's electrical energy. They are also the main source of greenhouse gas emissions. Reducing the use of fossil fuels for electrical power generation will have many benefits like reducing environmental pollution and improving health in societies. The increasing use of renewable energy sources for electrical power generation has been extremely helpful in reducing carbon emissions. Renewable energy sources must also be economic and generate electrical energy at a lower cost than conventional fossil fuel sources. Electrification of industrial processes and transportation technologies will reduce the dependence on fossil fuels and help in minimizing environmental pollution.

REVIEW QUESTIONS

10.1 Describe the environmental impacts of electrical energy generation using fossil fuels.

10.2 Describe the environmental impacts of electrical energy generation using nuclear fuels.

10.3 Describe the renewable energy and the main renewable energy sources.

10.4 Describe the generation of electrical energy from concentrating solar power technologies and from solar photovoltaic systems.

10.5 Describe how solar photovoltaic panels are oriented for optimum performance.

10.6 What is grid-connected photovoltaic system?
10.7 Describe the generation of electrical energy from wind energy systems.
10.8 Briefly describe the hydroelectric power generation systems.
10.9 Briefly describe the generation of electrical energy in nuclear power plants.
10.10 Describe the generation of electrical energy from geothermal energy sources.
10.11 Describe the generation of electrical energy from wave and tidal energy sources.
10.12 Describe the generation of electrical energy from biomass and biofuels.
10.13 Write a short note on renewable energy economics.
10.14 Write a short note on the benefits of electrification of transportation systems.

References

1. Alexander, C. K. and Sadiku, M., 2017, *Fundamentals of Electric Circuits*, 6th Edition, McGraw Hill.
2. Bergen, A. R. and Vittal, V., 1999, *Power System Analysis*, 2nd Edition, Prentice Hall.
3. Blackburn, J. L., 1993, *Symmetrical Components for Power Systems Engineering*, CRC Press.
4. Breeze, P., 2014, *Power Generation Technologies*, 2nd Edition, Elsevier.
5. Cathey, J. J., 2001, *Electric Machines Analysis and Design Applying MATLAB*, McGraw Hill.
6. Chapman, S. J., 2002, *Electric Machinery and Power System Fundamentals*, McGraw Hill Higher Education.
7. Chapman, S. J., 2005, *Electric Machinery Fundamentals*, 4th Edition, McGraw Hill Higher Education.
8. Del Toro, V., 1985, *Electric Machines and Power Systems*, Prentice Hall.
9. Del Toro, V., 1992, *Electric Power Systems*, Prentice Hall.
10. Edminister, J. A., 1986, *Schaum's Outline of Theory and Problems of Electric Circuits*, 2nd Edition, McGraw Hill.
11. Elgerd, O. I., 1982, *Electrical Energy Systems Theory: An Introduction*, 2nd Edition, McGraw Hill.
12. Elgerd, O. I. and Van der Puije, P. D., 1998, *Electric Power Engineering*, 2nd Edition, Chapman and Hall.
13. El-Hawary, M. E., 1995, *Electric Power Systems Design and Analysis*, John Wiley & Sons.
14. El-Hawary, M. E., 2017, *Electrical Energy Systems*, 2nd Edition, CRC Press.
15. Faulkenberry, L. M. and Coffer, W., 1996, *Electric Power Distribution and Transmission*, Prentice Hall.
16. Glover, J. D., Sarma, M. and Overbye, T. J., 2012, *Power System Analysis and Design*, 5th Edition, Cengage Learning.
17. Gönen, T., 1988, *Modern Power System Analysis*, John Wiley & Sons.
18. Gönen, T., 1988, *Electric Power Transmission System Engineering Analysis and Design*, John Wiley & Sons.
19. Grewal, B. S., 1994, *Numerical Methods in Engineering and Science*, 2nd Edition, Khanna Publishers.
20. Grewal, B. S., 1996, *Higher Engineering Mathematics*, 33rd Edition, Khanna Publishers.
21. Gross, C. A., 1986, *Power System Analysis*, 2nd Edition, John Wiley and Sons.
22. Gross, C. A., 2006, *Electric Machines*, CRC Press.
23. Gupta, B. R., 2009, *Generation of Electrical Energy*, 6th Edition, S. Chand and Company.
24. Gussow, M., 1983, *Schaum's Outline of Theory and Problems of Basic Electricity*, McGraw Hill.
25. Hughes, E., Smith, I. M., Hiley, J. and Brown, K., 2001, *Electrical and Electronic Technology*, 8th Edition, Pearson Education.
26. Kaiser, J., 1998, *Electrical Power: Motors, Controls, Generators, Transformers*, Goodheart-Willcox Publishing Company.
27. Khan, S. U. and Nakhabov, A. V., 2020, *Nuclear Reactor Technology Development and Utilization*, Elsevier Science Publishers.

28. Kothari, D. P. and Nagrath, I. J., 2010, *Basic Electrical Engineering*, 3rd Edition, Tata McGraw Hill Publishing Company.
29. Kothari, D. P. and Nagrath, I. J., 2010, *Electric Machines*, 4th Edition, Tata McGraw Hill Publishing Company.
30. Kothari, D. P. and Nagrath, I. J., 2003, *Modern Power System Analysis*, 3rd Edition, Tata McGraw Hill Publishing Company.
31. Masters, G. M., 2004, *Renewable and Efficient Electric Power Systems*, John Wiley & Sons.
32. McPherson, G., 1981, *An Introduction to Electric Machines and Transformers*, John Wiley & Sons.
33. Mehta, V. K. and Mehta, R., 2005, *Principles of Power System*, 5th Edition, S. Chand and Company.
34. Meier, A. V., 2006, *Electric Power Systems: A Conceptual Introduction*, John Wiley & Sons.
35. Morley, L. A., 1990, *Mine Power Systems*, Information Circular 9258, US Bureau of Mines.
36. Nasar, S. A., 1981, *Schaum's Outline of Theory and Problems of Electric Machines and Electromechanics*, McGraw Hill.
37. Nasar, S. A., 1990, *Schaum's Outline of Theory and Problems of Electric Power Systems*, McGraw Hill.
38. Nasar, S. A. and Trutt, F. C., 1998, *Electric Power Systems*, CRC Press.
39. Nelson, V., 2011, *Introduction to Renewable Energy*, 1st Edition, CRC Press.
40. O'Malley, J., 1992, *Schaum's Outline of Theory and Problems of Basic Circuit Analysis*, McGraw Hill.
41. Pimentel, D. and Pimentel, M. H., 2008, *Food, Energy and Society*, 3rd Edition, CRC Press.
42. Quaschning, V., 2005, *Understanding Renewable Energy Systems*, Earthscan Verlag Publishers.
43. Rizzoni, G., 2009, *Fundamentals of Electrical Engineering*, McGraw Hill Higher Education.
44. Saadat, H., 2010, *Power System Analysis*, 3rd Edition, PSA Publishing.
45. Slemon, G. R., 1992, *Electric Machines and Drives*, Addison-Wesley Publishing Company.
46. Stevenson, W. D., Jr., 1982, *Elements of Power System Analysis*, 4th Edition, McGraw Hill Book Company.
47. Twidell, J. and Weir, T., 2006, *Renewable Energy Resources*, 2nd Edition, Taylor and Francis Publishers.
48. Weedy, B. M. and Cory, B. J., 1998, *Electric Power Systems*, 4th Edition, John Wiley & Sons.
49. Weisman, J. and Eckart, R., 1985, *Modern Power Plant Engineering*, Prentice Hall.
50. Wildi, T., 2002, *Electric Machines, Drives, and Power Systems*, 5th Edition, Prentice Hall.

WEBSITES

I. Anon., "Introductory Chapter – Aspects of Renewable Hydroelectric Power Generation", https://www.intechopen.com/chapters/55231, Last accessed on 1st March 2024.
II. Anon., "Phasor Diagram of Synchronous Motor", https://www.eeeguide.com/phasor-diagram-of-synchronous-motor/, Last accessed on 1st March 2024.
III. Anon., "Protective Relay", https://www.eeeguide.com/protective-relay/, Last accessed on 1st March 2024.

IV. Anon., "Oil Circuit Breaker Diagram", https://www.eeeguide.com/oil-circuit-breaker-diagram/, Last accessed on 1st March 2024.

V. Anon., "Busbar Protection", https://www.eeeguide.com/busbar-protection/, Last accessed on 1st March 2024.

VI. Anon., "Symmetrical Fault Analysis", https://webstor.srmist.edu.in/web_assets/srm_mainsite/files/2018/symmetrical_fault_analysis.pdf, Last accessed on 1st March 2024.

VII. Anon., "Sequence Networks and Unsymmetrical Faults Analysis", https://webstor.srmist.edu.in/web_assets/srm_mainsite/files/files/Chapter4.pdf, Last accessed on 1st March 2024.

VIII. Anon., "Electricity Explained: Electricity and the Environment", https://www.eia.gov/energyexplained/electricity/electricity-and-the-environment.php, Last accessed on 1st March 2024.

IX. Anon., "What Is the Energy Source of the Future?", https://science.howstuffworks.com/environmental/energy/energy-source-future.htm, Last accessed on 1st March 2024.

X. Anon., "The Future of Electricity: Electrification", https://www.energysage.com/electricity/future-of-electricity-electrification/, Last accessed on 1st March 2024.

XI. Anon., "Grid Connected PV Systems", https://www.alternative-energy-tutorials.com/solar-power/grid-connected-pv-system.html, Last accessed on 1st March 2024.

XII. Anon., "Solar Explained: Photovoltaics and Electricity", https://www.eia.gov/energyexplained/solar/photovoltaics-and-electricity.php, Last accessed on 1st March 2024.

XIII. Anon., "Solar Panel Orientation", https://energyeducation.ca/encyclopedia/Solar_panel_orientation, Last accessed on 1st March 2024.

XIV. Anon., "Solar Photovoltaic Energy Conversion", https://www.desware.net/sample-chapters/d06/d10-014.pdf, Last accessed on 1st March 2024.

XV. Anon., "Hydropower Explained: Wave Power", https://www.eia.gov/energyexplained/hydropower/wave-power.php, Last accessed on 1st March 2024.

XVI. Electrical4U, "Thermal Power Generation Plant", https://www.electrical4u.com/thermal-power-generation-plant-or-thermal-power-station/, Last accessed on 1st March 2024.

XVII. Electrical4U, "Schematic Diagram of Gas Turbine Power Plant", https://www.electrical4u.com/schematic-diagram-of-gas-turbine-power-plant/, Last accessed on 1st March 2024.

XVIII. Electrical4U, "Long Transmission Line", https://www.electrical4u.com/long-transmission-line/, Last accessed on 1st March 2024.

Index

For Product Safety Concerns and Information please contact our EU representative GPSR@taylorandfrancis.com Taylor & Francis Verlag GmbH, Kaufingerstraße 24, 80331 München, Germany

Batch number: 10397790

Printed by Printforce, the Netherlands